Renewable Energy-
Following
the Money

CRAIG SHIELDS

Table of Contents

Introduction

The concept of "following the money" is by no means new. Socrates told us 2500 years ago that "all wars are about money," a notion that's certainly as true today as it was in ancient Athens. Much more recently, we recall that the truth behind the Watergate break-in that brought down Richard Nixon and the hint the mysterious informant provided: "follow the money." The phrase has become so iconographic in our times that it has its own page on Wikipedia.

It comes as news to no one that the energy industry is the most profitable on Earth. Based on their rise to prominence in the 20th Century, the annual earnings of the top five oil companies alone (ExxonMobil, BP, Shell, Conoco Philips, and Chevron) are now measured in the hundreds of billions of dollars. And, though I'm not privy to their board meetings, it seems a reasonable bet that they intend to use that muscle to retain their position of leadership through the 21st Century. By all accounts, any attempt to alter these people's plans means fighting against the single most powerful political force in the known universe.

Of course, as the world's concerns over the health and environmental effects of consuming fossil fuels grow with each passing year, this becomes a much more interesting subject than it was a few decades ago. We live on a planet with about 200 sovereign countries, representing a wide range of types of governments, levels of affluence, and access to energy resources, not to mention citizenries with varying sensibilities regarding environmentalism and levels of interest in the subject.

Where is all of this going? I'm not sure. Let's follow the money, and see where it takes us.

This book is a follow-on to my first two projects: Renewable Energy—Facts and Fantasies, published in 2010, a survey of the technology, the economics, and the politics associated with clean energy, and Is Renewable Really Doable?, published in 2012, a "deeper dive" into the issues preventing us as a civilization from migrating to a safe and sane energy policy. Here in Following the Money, I present another set of interviews, these associated with the effects that economics and financial power have on the course of the energy industry; after the interviews, I offer a group of short essays on the subject of energy policy and how I believe we need to navigate the waters associated with a migration to a sustainable course in energy.

At the risk of giving anything away, I'm not promising a surprise ending. As I said in my introduction to my last book, those who believe that the working of the world have even the remotest bearing on what's good for the seven billion people who live here are in for a rude shock. What's going on around us, of which our world energy policy is only one example, is anything but an accident; it's a careful unfolding of events designed to further enrich an extremely slender minority of interests – and guess what? Change of any type – especially the radical change that our civilization so desperately needs in its approach to transportation and energy – is not part of their plan.

That, of course, doesn't mean that the exercise is of academic importance only. There is, in fact, a chance that humankind can forge a sustainable path to its consumption of resources here on Earth, but, if we are to make it happen, it will be because of a largely informed electorate – of people who understand and honestly care what's going on around them.

OK, seatbelts fastened? Let's follow the money.

Dennis McGinn—
American Council on Renewable Energy

I met Dennis McGinn immediately after the first presentation he made when he took the helm of the American Council on Renewable Energy (ACORE) a few years ago, where I recognized instantly that he was perfectly suited to the incredibly important and difficult task of driving clean energy into prominence. Not only are his skillset and personality perfect for the job, his background couldn't be any stronger; "Denny" served 35 years in the U.S. Navy before he retired as a Vice Admiral. He's a widely recognized energy and national security expert, as well as a regular participant in public forums about energy and national security, having been published in newspaper articles and opinion pages across the country.

Craig Shields: Thanks so much for making yourself available for this interview, Denny. Let's jump right into this. As I mentioned, the topic is "following the money," and I'm trying to treat this as justly as possible, but trying to be fair in a politicized world isn't a piece of cake.

I just saw a debate on Fox News covering biofuels, and one guy said, 'What you have here is essentially Leftists pushing anti-capitalists' clean energy down our throats using the prestige of the United States Navy to do it.' Those are some pretty harsh words. How do you respond to something like that?

Dennis McGinn: Well, you say that it's all about improving the Navy's mission effectiveness and strategy. I did an update on the Navy's biofuels that was in the Hill newspaper recently. I was on NPR the week before on the On Point show and basically they had a Cato person, they had Phyllis Cuttino from Pew and I talk about this whole DoD biofuels issue. Their belief is that anybody in uniform who has anything good to say about biofuels or clean energy is a puppet of a socialist plot. All of that stuff is just so bogus. It's incredible. They go searching diligently for anybody in a uniform, or retired, that has anything critical to say about renewable energy and they say 'See? See? I told you.'

It's just crazy. But anyhow, I think that there are lots of people on both sides with viewpoints that underwrite the fact that what the DoD is doing is exactly on the mark.

CS: Okay, good. What I wrote about it, for what it's worth, in response to Inhofe, the Republican in Oklahoma, is essentially this: The Navy needs to defend our country this week, but also 30 years from now. They need to look at this challenge strategically. You're telling them they can't? You presume to limit their ability to do their job?

DM: In questioning the Navy and DoD investment on biofuels, Inhofe's got a very strong view based on a couple of his staff. The Navy has somebody telling me that every dollar spent on anything like this does not produce the same military capabilities and strategic value that other things do. I suppose you can argue this point, but it just doesn't hold water when you really look at what the Navy is actually doing.

CS: Okay. Can you articulate that in a couple of sentences?

DM: I would say something along the lines of congratulating the Navy's leadership for their strategic vision; looking beyond the next quarter's earnings statement or the next election cycle and thinking down range about what the strategic environment is and what the challenges will be. Clearly one of the challenges is going to be a very, very competitive dynamic for fossil fuels.

They want to increase the size of their energy portfolio to power the fleet and all the other things and they are looking at all kinds of alternatives. They start a program with energy efficiency to squeeze the most military value, the most operational effectiveness of every gallon of fuel.

But then they also want to understand the choices that are available and they are looking at biofuels.

I would then ask: Is this new? No. The Navy used to be powered by nothing but wind and they started to go to steam using these infernal contraptions called steam engines. There was a big uproar. 'This is foolishness! What's wrong with using wind power for all these fine sailing vessels as has been done for over 100 years?' The Navy persisted. They brought steam to power the fleet and they used primarily wood. Then they decided they needed to establish coaling stations because the energy density of coal is higher than wood.

Everybody said 'This is craziness. This is foolishness. Coal is more expensive than wood.' But the Navy persisted. They did it for strategic reasons thinking into the future. Then they decided 'You know, we've got something that's even more transportable than coal and viscous and that's oil.' Then everybody was complaining 'This is more Navy foolishness. Why would they go away from coal? It's a proven commodity that's available. You have to assume the costs that are associated with oil.' And thank God they did. Then you fast forward to the late 1940's, the '50s, and more Navy foolishness on energy as they developed nuclear power.

This is what's been going on for a long time. Thank God for the Navy and all of its services and their ability to look strategically ahead and make the right kind of investments.

They are not going to put every dollar on this stuff. They did the research and development, the early prototyping and the operational testing to make this viable. The big complaint right now is, 'How can you spend $26 dollars a gallon for this biofuel stuff?' The big thing they need to say is 'We're not going to go to war with a Navy that powers itself with $26-a-gallon biofuels.' We're buying it now because it's new; we're testing it; we want to scale it up, and the prices will come down. We're not going to go wholesale off of petroleum right now because we can't. It doesn't make any sense. But we have to start. That's the essence of the logic.

CS: Well, it certainly makes sense to me and thank you for the historic background on that. As long as we're on fuels, is there any interest in synthetic fuels? One of the ideas that I've come across is a guy who has cracked the code with respect to how to synthesize high-grade hydro-carbons: high-octane gasoline, high-grade diesel, using off-peak wind—or off-peak anything, really.

DM: Yeah, there are a bunch of, I would say, early-stage ways of doing that type of thing. I think that they aren't right around the corner but they're certainly worth looking into. The thing that has also been looked into is using Fischer-Tropsch for coal to liquids. But if you take a look at the energy equation and the greenhouse gas impact, it's nuts; it's not a good thing to do. I'm glad they are not pursuing that.

CS: Well, it turns out that Fischer-Tropsch is the last of five chemical processes that ultimately synthesizes the hydrocarbon but it's not coal to liquids; it starts with the electrolysis of water, and taking waste CO_2 from concrete plants and coal-fired power plants.

DM: Fischer-Tropsch, in and of itself, is a very viable process. The question is how you use it. Do you use it with the right kind of feedstock produced in a sustainable way like with the electrolysis of water produced from wind or solar? Hey, that's fine. But coal or gas to liquid isn't viable.

CS: I completely agree.

I know I've mentioned this to you before, but I applaud your unflappable temperament; there is lunacy all over the place, and instead of getting angry at it, you calmly point it out for what it is and move on. I think that's extremely positive. That's probably why you're the president of ACORE and I'm not.

DM: [chuckles] Well, I'll tell you what, there is an absolutely critical role for passion, and I occasionally get worked up. Sometimes it doesn't manifest itself externally but I care very, very deeply about these issues.

CS:Yes, I know you do. I know you're not indifferent to them. My point is that you conduct yourself as a professional and I think that's important.

Maybe we could end off on the previous subject, just summarizing. I know that it's different in Europe and Asia, but here in the United States we have this kind of bifurcation along party lines where it seems that the right wing, for whatever reason, denies the importance of sustainability and cleantech across the board. I.e., they say that climate change mitigation and renewable energy are unnecessary, that they're too expensive, that they're part of a leftist plot, etc. Can you comment on that for me, please?

DM: Sure. There is basically a lot of politics especially at the fringes, if you will, at the ends of the spectrum on both sides. On the Republican conservative side I think there is a certain aspect of 'If Al Gore liked it, I hate it and it can't be true. If President Obama is pursuing policies that in some way would deal with the threat of climate change, then I have to oppose them because I'm opposed to everything that he does.'

That's the ugliness of the politics. I think you'll find more centered, rational people who see the business case and the environmental case. I always use the expression 'The root word of "conservative" is "conserve."' In some of these political arguments, people lose sight of what those core American values are. A lot of people beat their chests and say how patriotic they are and how they love America— American exceptionalism and all that stuff. Well, all of that is based on a set of historical values that are absolutely critical to where we got to, and where we are, and we've lost sight of that. Thrift, self-reliance, all of those things have to do with what we're trying to do with energy efficiency and renewable energy.

CS: Tie that together for me, please, Denny, in terms of thrift and self-reliance. I guess you're talking about distributed generation?

DM: Yes, that's a key part of it. I mean a lot of people raise all kinds of flags about increasing utility bills and letting the power utility company do this or do that. There's a tremendous amount of push-back about any kind of power outage, understandably. In some cases it goes beyond inconvenient; it's life-threatening in some rare cases. I think that the people like this idea of having a little more control over what they pay for electricity, and energy efficiency is a nice place to start. And, in terms of self-reliance, some of the most conservative people would love it if they could be off the grid and independent of large utility companies.

CS: Yes, that's a good point. I was a marketing guy before I was a renewable energy author, and I've always said that if I were trying to sell this to the American people it would really be on the basis of patriotism. If you really love this country, the continued dependence on fossil fuels has only downside, especially militarily. I interviewed James Woolsey for my first book and he has me pretty well convinced that from a national security perspective this is a true disaster, this reliance on fossil fuels.

DM: Jim and I have worked very closely together over the years in this area. I fully concur. I guess the sound-bite that captures it for me is that the United States faces energy security challenges, economic security challenges, and environmental security challenges that are local, regional, and by the way, global. All of those can be viewed through the lens of national security whether it's military, diplomatic or economic. If we are really serious about preserving our security and our prosperity, we really need to get serious about how we address those economic, environmental, and energy challenges in a way that does not cause harm in any one of those dimensions, but rather produces a positive, synergistic result. That's why I'm in renewable energy because I see renewable energy as being good for the economy; it's good for energy security; it provides a more diversified portfolio of energy choices, both for electricity production as well as for transportation fuel. And it is so much better for the environment.

Obviously there is the great theological debate about climate change that you just mentioned but people tend to personalize environmental security when you start talking about local and regional environmental impacts as a direct result of the ways in which we generate electricity or produce transportation fuel.

CS: Very good. Once we've agreed that this is a good thing, and I think we have, let's talk about how to get from here to there. First, let me put a couple of things on the table. There are externalities of fossil fuels which we systematically ignore. We also subsidize fossil fuels at least two or three times the degree to which we subsidize renewable energy, which creates a playing field that's tilted away from renewable energy. Could you please talk about the playing field—where it is and what effects that creates?

DM: Sure. Before I do that let me go back to conservative America for a moment. Craig, I strongly recommend that you get a hold of a former South Carolina Congressman, Bob Inglis. He was "tea partied" out in 2010 because he had the audacity as a tremendously social and fiscal conservative to actually acknowledge there was such a thing as climate change and that we had something to do with it. He came to that conclusion through a great deal of study and personal reflection.

He is an absolutely outstanding individual and has just started an effort that is affiliated with George Mason University here in Northern Virginia to make the case to conservatives, fiscal and social conservatives, for a solution set that addresses the challenges of energy security and environmental security. Basically it's all about 'Hey, let's get the free market rolling here.' Some of his solutions, and I'll segue into the level playing field, would in fact, in a revenue-neutral way, provide resources that would help to scale up renewable energy and would also capture the social costs, the externalities, of all of the elements of our energy portfolio.

CS: I've come across him before, in fact, I've written a few pieces on him. That kind of true bravery is rarely seen anymore.

DM: So, level playing field. There are a lot of differences here because, obviously, the oil and gas industry has been around for over a hundred years. They are well established in terms of infrastructure, markets, capital, brand, etc., all deeply embedded in society and the economy. In addition, there is a real cognitive dissonance here when people say on the one hand, 'Government shouldn't be involved in private sector energy' yet you see the amount of subsidies. In fact, they have been going on so long with oil and gas, I think of them as entitlements. They are just continuing to get those tax payer dollars, if you will, either through tax breaks or direct subsidies. That's okay, I guess, if we want to be in the business of subsidizing energy but we've got to recognize that we have, with renewable energy, just about across the board, a *growing* industry—one that has so much more potential for the creation of jobs and producing better outcomes than what our current portfolio of energy does with its over-reliance on fossil fuel.

There are a couple of ways of achieving this. One would be to subsidize renewable energy and energy efficiency at levels that the fossil fuel industry receives. Another, if you're concerned about the government getting out of the subsidies business, is to take them away from everybody but figure out a tax mechanism, a green carbon tax or something along those lines, revenue-neutral, that takes into consideration the externalities and social costs of each aspect of our energy. That would level the playing field, but you need both. You can't just say 'Okay, in the name of reducing the deficit we'll take away all subsidies.' That's fine but you also need to take into consideration the social costs.

CS: Yes. As I always say, these costs are simply being buried. In other words, they are being incurred but they are being paid by my grandchildren; we're just asking somebody else to pay the cost.

DM: I'll give you an example of that in a national security sense. We've got tens of thousands of young men and women in uniform deployed around the world, using hundreds of billions of dollars in military equipment across all the services. One of the major national rationales, not the only one certainly, is to ensure the free flow of the life blood of the global economy called oil.

If you start factoring in that portion of the defense budget, for example, or that portion of healthcare costs that are directly related to our over-reliance on fossil fuels we would be paying at least, at a minimum, double what we're paying at the pump for a gallon of gasoline.

I've seen analysis that captures this quite well; in fact, about two years ago, a retired general Chuck Wald and I—he's an Air Force guy—came across this analysis and basically it said that the cost of a gallon was far larger than what we're paying.

CS: Exactly. Now let's get to the other side of that, the development of renewable energy vis-à-vis an industry. It is true that oil and gas have a hundred year head start on renewables, and it is true that this has created certain barriers. Let's talk a little bit about those barriers and how you see a path opening up for clean energy to overcome them. Again, some people look at what I'm doing with renewable energy and say, 'That's a nice hobby.' Unfortunately, there is a grain of truth there, at least as far as perception is concerned. Can you comment on that please?

DM: It's a perception that is rapidly fading away as more and more people are touched in some ways, some visible, some less visible, by renewable energy. I'll give you an example. How many times in the past week have you been driving and seen a highway sign that has solar panels providing its information? Anyone who flies into and out of Denver International Airport, if they are looking in the right direction, would likely see a huge photovoltaic array that is helping to supply a good portion of the electricity that Denver International uses. We are using right now a ten percent blend of ethanol, a bio-based fuel, in all of our fuel. Can you imagine if we weren't doing that? We'd be using ten percent more fossil fuel.

There are ways in which the industry is growing. If you look over the past three to five years, the actual deployment of renewable energy, the big ones have been wind and solar. Measured in megawatts, it's huge. The price is coming down and because of that we're starting to get to an economy of scale. I think that a lot of people have this misperception that it's immature, or that it's elitist; there are all kinds of myths.

First of all, they say that renewable energy can't scale up; second, that renewable energy is too expensive. Third, renewable energy is totally dependent on government subsidies. Fourth, that it is a fad, a fashion, mostly adopted by elitist people, leftists, socialists, on the coasts of the United States. Yet you drive through Iowa, Texas, Oklahoma, and see the wind farms there and people say 'Oh, this is real.' Even West Virginia, on the ridges of Appalachia; it's starting to work its way into the American society and the American economy in ways that people are starting to understand 'Oh, this is real and it's got benefits. I like this.' So I think this is happening.

CS: I hope you're right. But let me challenge you there, simply referring to the title of the book, *Renewable Energy: Following the Money*. Going into this, my concept is that this will happen when the moneyed interests say it's going to happen. So, in other words, the fact that people have trillions of dollars of infrastructure sunk into oil and gas, not only in the United States but all over the world, means that nobody in their right mind is going to expect the most powerful people on Earth to just write off trillions of dollars' worth of investments. Not to mention the revenue streams that they have made. So how real do you believe the transition to renewables could be, given that it's competing against the most powerful interests on Earth?

DM: I am *assured* by a look at history where there were tremendously powerful interests before and they have been overtaken by market forces. Dick Foster from McKinsey Group more than 10 years ago wrote this book

Creative Destruction. There is a lot of that going on. If you do a side-by-side comparison of the real costs, the real benefits, and the real risks of each form of energy for both transportation as well as for electricity, renewable energy wins across the board. That is a market force that is starting to get traction.

You were up in New York (at the Renewable Energy Finance Forum). We didn't have some small VC out of Boulder, Colorado trying to green the world. We had Morgan Stanley and Credit Suisse and Deutsche Bank and Bank of America and CitiGroup and all of these real financial institutions, as well as some very, very well-funded venture capitalists and equity investors that are literally putting hundreds of millions of dollars – indeed, billions of dollars – into renewable energy. It's not as rapid as it would be if we had a more balanced set of state and federal policies that took a lot of the uncertainty away from it, but it is in fact happening.

I mean, the big thing there that captures the uncertainty is the scheduled expiration of production tax credit. We're seeing companies in the supply chain that produce wind turbines that are laying people off now because they can't have any assurance that the PTC is going to be extended— and if it is, for how long. The production tax credit is a tax policy that has produced a tremendous amount of jobs— in fact, net revenue for the U.S. government because of that job creation.

CS: Speaking on the Renewable Energy Finance Forum, one of the hot topics this year was Master Limited Partnerships. Do you want to cover that?

DM: Sure. One of the things that is really important is to level the playing field of project finance. The costs of a renewable energy project are made up of the usual components: materials, labor, service costs, analysis, law firms, bankers, fees, etc., and also the cost of capital. That cost of capital and the availability of capital is not what it is for other economic investments—certainly not for mining, or for oil and gas exploration.

Two Senators, Senator (Jerry) Coons from Delaware, a Democrat, and a Republican (Chris) Moran from Kansas, a couple of weeks ago introduced what they call the MLP Parity Act. This is something that we hope in the next Congress is passed into law. It will allow the mechanism of Master Limited Partnerships to bring together a lot more money from sources that currently can't invest in renewable energy using this mechanism. The result is a lower cost of capital -- many basis points lower.

Another thing that is happening is there is some effort inside of the Department of the Treasury to see what portions of renewable energy developments or investments could be made using the real estate investment trust (REIT) mechanism. This also would bring a much larger spectrum of investors into renewable energy and would make capital available at lower costs.

The bottom line, I guess, is we've got to level, not just the government mandate, government policy, government tax equity, government investment, loan guarantee field. That's certainly helpful but we have to also level the playing field for the availability of and the cost of capital to do these projects, because it's a dollar for dollar basis. You lower the cost of capital to do these projects, and you lower the cost of delivered renewable energy.

CS: Yes. Good point. Let me ask you about this. I first met Ray Lane, the managing partner of Kleiner Perkins, at one of the great auto shows; I guess it was Detroit, a couple of years ago. He was the last speaker and at the end of two days of listening to self-congratulatory people, talking about how we're putting Americans back to work and how we're regaining our former glory, he basically said 'That's not true. All this clap-trap about the great work we're doing in 21st century transportation and energy is disingenuous. America actually *did* a great job, in terms of leading the world in IT and communications and networking and the Internet. What we did in the late 20th Century was fantastic, but the concept that we're doing it again in energy in the 21st Century is completely untrue.'

It was a very sobering moment and it took a lot of guts to stand up in front of a couple thousand people who were slapping each other on the back, and to tell them simply to knock it off. I bring this up because it gets at my point that when you look at what's happening at kind of the macro level, you see China investing huge amounts of money. You see China taking this very, very seriously and the United States kind of content to bicker about it and let the rest of the world pass us by in terms of leadership. First of all, do I have that right or wrong? How would you comment on that?

DM: You've got it exactly right. Let's use a U.S. and China comparison for example. Most of the technology of renewable energy that the Chinese are developing and using domestically as well as exporting profitably was invented in the United States. What we have lacked is a sense of urgency, a sense of seriousness in using this for our own benefit. It has been because the vested interests in energy—read coal, oil, and gas—have made us pretty complacent in terms of 'Gee, the price is nice and cheap. It's available. Yeah, there are occasional crises that drive the price of oil up but it will come back down.' We've basically been asleep at the wheel in this regard.

China, on the other hand, has a rapidly growing economy that has to rely on more and more sources of energy of all sorts: a Three Gorges dam, those one-a-week coal plants, the rapid investment and deployment of renewable energy. We've got all the necessary ingredients in the United States still to lead the world in the development, deployment, and economic benefits of renewable energy. Not in a dominant way, or a zero-sum way, but in a very win-win type of situation for us and all of the regions of the world. What we have lacked is a viable, long-term, consistent energy strategy or energy policy. That is really, really hurting us.

I have talked with many renewable energy investors, especially VC's who have said, 'I am really, really, bullish about the future of renewable energy globally. I am bearish about the future of renewable energy in the United States.' Because of that lack of policy, consistency, short-term political bickering, we are losing an opportunity every day that goes by that we don't get our act together.

I'm not talking policies that throw huge sums of money at this, like the Chinese. I'm talking policies that would create market certainty for developers of technology, for deplorers of technology, for investors to get the hundreds of billions of dollars of capital that are waiting on the side lines looking for a good investment. We could do that.

CS: Indeed.

DM: There is so much at stake here. To give you a specific, I have talked with folks and visited in China and heard directly from the managers of photovoltaic companies in China that are actually selling below their marginal costs because of the great pressure from the government.

You talk about government subsidies? My God, just look at the unbelievable amounts of money that the Chinese have put in. I'll stick with the photovoltaic business and say that there has been some good analysis, if you look at the price of labor and you look at the price of materials in the United States, in Europe, in China, for the production of photovoltaic capability; it's in the ballpark. China is getting a little bit more expensive, by the way, as their middle class starts developing a little bit, but it's pretty much on par; you can work in a competitive field like that.

Add to it the cost of capital and the Chinese win hands down. In fact, there isn't really a cost of capital for Chinese photovoltaic manufacturers because there is so much government subsidy. That's why you can't compete in many ways with the actual manufacturer of photovoltaics in places outside of China. Now there are some good stories about taking innovative, whole value chain approaches to getting photovoltaics deployed in the United States, for example. There are some good innovations in terms of developing a much more efficient conversion from light to electricity. There is some good innovation in terms of the form factor in which it's deployed. It's not like we're done for, but every day that goes by we're missing the boat in terms of a potential leadership position. You want to follow the money? You have to have the right kind of balance between three critical factors: technology, finance, and policy.

CS: Wow, terrific. Anything else you'd like to add?

DM: I will close the conversation by saying that I am optimistic, though not unrealistically so. But I know that because of the benefits, as compared to the costs and the risks, this is going to be a key part of our energy future. I know I come across as a zero-sum guy, but I recognize that the United States has benefited tremendously from the efforts of the folks in fossil fuels. Having said that, there is a time when, for a whole bunch of reasons, both internal and external, that you need to start down a different path. Not in a stupid way, not in a radical way, but in a rational way.

You do need to balance the playing field, and you do need to expect that the people who are stakeholders in the value chain are going to push back. All the legislation around DDT, lead-based paint, asbestos, all the of the safety features we have in our automobiles, was opposed by people in the value chain.

CS: Right. The only reason we have seat-belts in our cars is that someone outside the auto industry made it happen.

DM: Yes. Despite the protestation of the industry: people don't want this, they won't buy this, it will ruin the industry — and, by the way, we're already working on the solution, and we'll have it for you in 10 years.

CS: Ha! You're absolutely right. I have people in my life who say that I'm trying to fix something that isn't broken. When we turn on our lights, they come on without fail. We have cheap, reliable energy. And I have to admit that they have a point. The issue, however, lies in what you mean by "broken."

DM: I couldn't agree more. Let's talk again soon.

Brian Wynne—
Electric Drive Transportation Association

B rian Wynne is the president of the Electric Drive
Transportation Association, an international member-
based organization, promoting battery, hybrid, plug-in
hybrid and fuel cell electric vehicles and infrastructure.
I've known Brian for about five years; in fact, the tran-
script of my interview with him in late 2009 functioned
as my chapter on electric vehicles in my first book. I've always admired his passion, as well as his accessible "plain-
talk" manner of explaining his position on what can be a
difficult and controversial subject.

Craig Shields: Brian, you and I go back several years.
In fact, I had known you for a while before I interviewed
you for "Renewable Energy – Facts and Fantasies," and
that was three years ago. Perhaps we could start with an

update on the Electric Drive Transportation Association – its mission, its successes, etc.

Brian Wynne: We continue to advocate as we are an advocacy group. We love it here in Washington and we'd like to think of ourselves as an organization that has punched way above its weight class. The truth of the matter is that we had a very nice tailwind for a while and we still have a lot of support for what we are doing. And as soon as gas prices start getting unreasonable again, which they will very soon, we will be back on that particular message.

The advocacy piece has gotten more complex because we are in the process of operationalizing electric transportation, which means that all manner of regulatory things are now starting to come into play -- everything from safety regulations around the shipment of lithium-ion batteries on airplanes to e-MPG and what's on the stickers for the cars.

Of course, all the work that we have done has created a good core of journalists that write about this. Not just environmental journalists but auto writers, technology writers, business people magazine writers and the life-style magazines like GQ and Esquire all jump in once in a while. But recent developments have just demonstrated how distractible the media has become.

CS: In other words is there a sense in which they have turned against you?

BW: Well you are reading the same stuff I am. It's gone from the sun rises and sets on this technology to complete nonsense about what's happening now.

CS: For example it's a failure, because we only sold 17,000 units.

BW: I am sure that is a spit in the ocean to the number of other vehicles we sell, therefore this can't possibly work. This looks right past all the hybrids that have been sold and the number of models that are available at multiple price points. Let's ignore the fact that transit buses are now mainstream already. There is a tendency to focus on something like a Volt fire which suddenly becomes "combustible cars." Excuse me; it's called the "combustion engine" that is what we are trying to replace.

CS: That is exactly right.

BW: If you were writing a story about car fires you would really want to be focusing on.....

CS: the six people a day who die in gasoline fires.

BW: I hadn't heard that statistic, but yes, and that's even more disturbing. I have seen anywhere from 235,000 first-responder calls to the 250,000 vehicles that burn on our roads every year. But we have one Volt catch on fire and it takes months to figure out what happened. They spend weeks trying to replicate it and after they have crash-tested severely abused battery packs, and gotten one to spark and flame that becomes "Volt's fires." .

One can get pretty worked up over these things, but I rewind to the point where we were reading and trying to get people educated about the challenge that we are up against here, which is to transform how we move people and goods in this country; that's really the game. The game is to move off fossil fuels as quickly as possible.

Sorry to have gotten worked up, but let me close by answering your question as to what we do here. I need to keep my emotions out of this, and be very dispassionate and very fact-based, because the story is actually a very good one. All of this is going extremely well in my view and we have the numbers to back that up. Craig, you can help, as we need to educate, not just opinion leaders through media, but we need to educate the consumers. That's a huge job.

CS: Good point; I'll do what I can.

Let's talk about the EV platform generally. I think most people have a sense of the fact that we in the United States are borrowing an incremental half-billion dollars a day and sending it to people who hate our guts -- and they don't like that. In addition to the national security stuff, they recognize various environmental and health-related issues about fossils fuels generally. Maybe we could start with the source and cost of oil.

BW: It doesn't make any difference where it comes from as we don't control the prices, that is controlled by OPEC. OPEC just turned over the chairmanship from Iran to Iraq. Do I need to say anything else about that? This is a huge risk. It's February but even at the end of January we were hovering around $100 a barrel. We have the weakest (oil reserves) in history, and the global economy is starting to turn. You look at the volatility of this price, and it's scary. I am an economist by training, if you look at the volatility of this commodity it is rivaled by few others.

CS: Having said that, electric transportation in the United States means that the fuel mainly comes from coal. If you believe that you are going to charge EVs at night, you are charging with whatever is available at night which is largely coal. Therefore, there are people who wonder about the well-to-wheels analysis if you could speak to that?

BW The well-to-wheels analyses we have demonstrate that it is better to plug your car in than to use gasoline even at today's generation mix. With more and more natural gas replacing coal and additional renewable coming on-line, that picture just gets better in the future.

CS: So we are definitely going in the direction of natural gas and who knows about renewable energy?

BW: The point is that the generation mix is going to change over time. We will be less coal-dependent; it doesn't matter even if we are 50% coal generation right now. The overall big picture is that it is still better to plug your car into the grid. To your point about people plugging in at night, I look at a place like Texas, which is of course the backyard of the oil and gas industry. They have built so much wind generation capacity in West Texas that their energy prices could go negative when the wind blows at night. Now they have still got some transmission to build and they are in the process of doing that, but they have recognized that most people's cars are sitting in their driveways or in their garages at night when all of this energy is available.

CS: Yes, if everyone lived in West Texas we wouldn't be having a problem with this. But the population centers are 1500 miles from West Texas.

BW I am not an expert on this but I sit on the DoE Advisory Committee on Electricity and there are a lot of people trying to figure out, absent a climate policy in this country, what do we do with all the coal plants? Give us a policy signal because we are going to have to retire a bunch of coal plants: win, lose or draw. They are coming to the end of their useful life and we recognize that building more coal plants (is not a good idea), as there are carbon challenges with them.

But the utilities actually make money on capital investment. That is how they make money for the most part. They have to create electricity somewhere by a capital investment in a plant that they then amortize; If they don't have a plant to amortize then they don't have electricity. So the business model for most utilities is essentially a capital investment plan.

Most of the heads of utilities in this country are bankers. The reason for that is they are essentially making long-term bets based on what's happening with electricity in the future. The first thing to be aware of is that they are looking at a lot of generation coming offline, and being retired, and they have to replace it with something. Also, there is not a lot of load-growth out there; we have a shift to greater conservation. Pacific Northwest National Labs did a study that demonstrated how much energy is available at night. Given the capital investment that we have already made nationwide, we could fuel 73% of the existing light-duty fleet with off-peak kilowatt hours. That's coming through in our electricity rates. We have already invested in that capital, some of which is going to come offline. That is like investing in a bunch of airliners and then having them sit over at National Airport.

CS: So we're not literally dumping off-peak power back to ground, but there is idle capacity?

BW: There is some going up the smoke stack; I will come back to that piece too. There is an even better story there.

CS: I understand that there is a limit to which you can tamp coal plants; in fact, I have looked in vain to try and figure out how much power we are literally dumping every night.

BW: Well let's talk about the capacity we are just not using.

CS: You say, "capacity we are not using," but that means "coal we are not burning." If you put more load on the grid at 3:00 AM, you are going to be stoking coal plants.

BW: Let me back up one step and let's make sure we are trying to solve the same problems. There are multiple challenges that we can address with electric vehicle technology and that is part of what I think people need to be educated about.

We started with how much gas are we using and why is that a problem. Just a quick statistic, $448 billion was spent by the consumer on gasoline last year alone in 2010, that was $100 billion more than in 2009. That's money that could have gone back into the economy. 60% of that goes overseas. So just for moving people and goods in this country we have the opportunity to utilize something that we have already invested in and we are paying for with our electricity rates, use it or not use it.

We will come back to feedstock in a minute, but for now, just realize that you have to pay for those plants, and you are passing that amortization through in electricity rates. We are paying for that investment which is going to come offline, so we need to invest in something different. Meanwhile there is no electricity load-growth to pay for that, which means that we are looking at electricity prices going up; it's just math.

We have already agreed that gasoline is bad for the environment and for the country. We are vulnerable in terms of energy security. So there is a natural agreement here that we want to shift from one commodity, which is a global fungible commodity, and, because we don't control the price, it's very volatile. We want to move to one that is much more stable and available and fuelled from multiple sources. We need to make decisions about those sources in the very near future, for investments that are going to last for the next 40 or 50 years: nuclear, wind, solar, natural gas, coal, geothermal, hydro, etc. Multiple feedstocks have come into this generation mix and we need to make some decisions about that generation mix.

But one way or the other you are looking at capital investment. So what are you using to pay for that, when there is no load-growth, and it's coming through in higher electricity rates? This is part of what is little understood: the more cars we plug into the grid, the lower our electricity rates are going to be. I may be over simplifying it but I was on a call yesterday with some folks that are doing the back-end stuff. It is very similar to what Willett Kempton has done at the University of Delaware; (U.S. Department of Energy's Federal Energy Regulatory Commission) FERC Chairman (Jon) Wellinghoff talks about this all the time. You need to keep a certain amount of spinning reserve at night in case there is a spike on the grid and you need to shift to it. So that's the dollars gone up the smokestack. Or you need to fire up the very inefficient peaker plant. This isn't my area of expertise; there is a lot about the utility industry I don't know. But what I do know is they need to keep a certain amount of spinning reserve.

Now imagine you have a bunch of cars plugged into the grid at night and you are using much more capacity; you don't need more spinning reserve. If I can identify where my cars are and there is a spike in the grid I can slow down the electricity to those cars or stop it. It will have no damaging impact on the batteries if you just stop and start again. Tons of dollars that are currently going up the smokestack are saved.

CS: So this contemplates smart grid, but not necessarily V2G (vehicle to grid)?

BW: Right. It absolutely is not V2G, which is taking energy back out of the batteries and using it as if it were distributed generation; I am not talking about that.

This is a smart grid application where you need to be able to identify an asset I am fuelling. I have a contract with a customer where the customer has sent me a signal that says I just need it full for this time. So there are opportunities here for more efficient energy use from domestic feedstock, that's all good.

CS: Now let me ask if you know who John Peterson is?

BW: No.

CS: He is a very smart and successful IPO attorney – an American who, when he became a gazillionaire, moved to Switzerland. In my estimation, he is one of the very few credible critics of electric transportation. What he points out is that we are going to run into resource issues especially with respect to nonferrous metals. As more consumers come online worldwide, as a huge population in Asia becomes more urbanized and consumer-oriented, they are all going to want appliances of various types. The average person consumes approximately seven kilograms a year of nonferrous metals. He says that electric transportation will cause the world to hit the wall in terms of getting all the stuff they need. Have you run into anything like this?

BW: No I don't know that argument. But if he is so smart and he is a good investor then he is probably thinking: How do I invest in that and make more money out of it? Because we are going to continually need to move goods and people in this country. I am happy to try and figure that problem out, because we have time. And it's no secret that if everybody wanted to buy an electric vehicle tomorrow we couldn't make them that fast. I am really worried about that.

CS: Well that's true you want supply and demand to mirror one another.

BW: I would like to have an orderly ramp and that would give us time to deal with any resource constraints that are out there. I am not saying there are no issues; I am asking how we address these issues as these vehicles come into the marketplace.

Let's not forget that there are enormous wings of the transportation world that must become electric drive. The Blue Water Navy is a classic example and of course submarines have been electrically driven for many years. They use batteries and generators to put energy back into the batteries; they run silent and they run deep. But now you have cruiser class battleships; what kind of fuel can we save by hybridizing these things? You have locomotives; GE makes hybrid locomotives which are very fuel-efficient. These are the kinds of things that you can drive with federal policy; it's low-hanging fruit.

Why are transit buses now 50% hybridized? Because the federal transit administration pays for a high percentage of that bus. It is federal policy driving us in a direction that will address the problem. We tend to focus on light-duty vehicles, and think about how many are going to be sold around the world, but it is a much bigger picture than that.

CS: As long as we're talking about political philosophy, the fact that the federal government can spend money that it doesn't have to try to make something happen is one thing, but is that really what you want in an ideal world?

BW: I would turn that argument around and say: Can the federal government afford not to do this? What are the externalities here, to use an economic term? What is the real cost of being completely dependent, which for all intents and purposes we are, on one global fungible commodity here. Let's add that up.

Now let's look at subsidies; the NRDC captured that at $78 billion over the next 4 years that the oil industry will get in tax credits alone, to do what? Exploration? Start looking at what it costs to defend these supply lines. We have got flat tops steaming through the Straits of Hormuz right now. We have seen this movie for many centuries. I am sure you read the book Turning Oil into Salt. We used to fight wars over salt and we don't do that since we invented refrigeration. We are fighting wars over oil now, but what is the cost of that?

We are at the intersection of energy, environment and transportation. These are three heavily regulated areas. There is going to be policy here. The question is where do we want to drive this thing? So it's more like the federal government cannot afford to do this. It is about priorities and I can't think of many things that can be more of a priority right now. We have got people dying over it.

CS: I know that you read some of the stuff I write, so you know that I cover this all the time.

BW: Yes, I know.

CS: OK, super. Moving on, a very clever guy once told me that the consumer has a vote in this too. Do you know who this clever guy is by the way?

BW: It is probably me. I have used that line a few times.

CS: Yes, it was. (laughter) So let's talk about where the consumer is with respect to an EV purchase. Before you answer the question let me tell you what I guess. He gets the $7,500 federal tax credit and perhaps some more from the state. So that's good, but let's look at the Ford Focus Electric from his eyes: it's $15,000 more in MSRP. So the payoff in terms of fuel savings, unless he drives the thing half a million miles, it's going to take forever to get his money back. And all the while, he has to plug it in and worry about range issues.

BW: I'm optimistic that this technology is going to take hold sooner rather than later, because when consumers get in the cars and actually drive them, they love them. When people walk into a showroom that are interested enough to come and *look* at the car, 35% of those people say they are interested in *buying* the car. After they have driven the car the percentage goes up into the 60s.

CS: They are cool; there's is no question about it.

BW: They are a dynamic experience. People buy cars for a number of reasons, but one thing that absolutely matches up with my own personal experience is that I enjoy driving again. I have driven SUVs, small sports cars – lots of different vehicles. I enjoy driving more now than I think I ever have because of the dynamic experience I am getting with my Volt right now. But I would get into an i-MiEV as I get the exact same experience.

CS: I love the i-MiEV.

BW: The i-MiEV after the tax credit is $21,000. So what I am optimistic about is that people buy vehicles for many reasons, and that suggests to me that our success will be a function of offering multiple types of vehicles at different price-points. We will need to educate the consumer so that they understand the differences between a pure battery EV, a plug-in hybrid, a range-extended EV and a fuel cell EV. They need to understand the differences between these and the different fuelling options.

By the way, if you don't like plugging your car in, inductive charging is going to change that for you. That technology is already becoming mainstream for cell phones. We have major players that are coming to our conference this weekend.

CS: The Qualcomm Halo thing?

BW: Yes. This really is like magic. You drive your car into your garage and it starts to fuel. You never have to go anywhere; you never have to plug it in. I started out talking about plugging my car in and I changed my story from "it will be really easy to plug your car in" to "it's really convenient to plug the car in, because I don't have to go to a gas station anymore."

I think of this from a responsibility standpoint and I am hoping that the utilities will get ahead of this and offer price signals that encourage people to charge at the right times; I think that will be important over the long haul as we get more buy-in. My car is charging right now downstairs; we have three chargers down there. But when I go home, I plug in with a set default to give me a full charge by 6 o'clock in the morning. So from my standpoint, it's increasing the awareness out there and having people step back and ask, "What do I really drive this car for? What really floats my boat about this car? What are the operating costs of this?" You may have seen the report recently from (I'll chase it down for you) -- the best operational costs is the Chevy Volt.

CS: Wow, that's interesting.

Given your position (as president of a membership organization whose sponsors include GM) there are probably things you don't want to say about the Chevy Volt. But I would think that plug-in hybrids or range-extended hybrids, which, in my mind are the same thing, are at best an interim step. To what degree do we need both an internal combustion engine and an electric drive? You are lugging about both of them whether you need them or not. If this has merit, it is only in the period of time before we have ubiquitous fast charging. Put another way, we are certainly not going to have plug-in hybrids by 2050. True?

BW: I don't know that fast charging is a goal, when there are technologies coming like the dynamic inductive charging built into the road.

CS: That's really going to happen? That's exciting.

BW: Yes, but it's further out there. That's like V2G and we don't have all the technologies and we haven't worked out all the business case for that, but there are now going to be live demonstration projects. Entire cities are going to be putting in for a demonstration. I don't know exactly where this goes, but what I do know is there are many vehicles in the fleet and so it is never going to be one size fits all. You look at UPS, which is buying a certain number of class 5 and 6 delivery trucks with a certain size battery pack in it. They have a plan for when that battery pack is done in that one, and it has a certain amount of capacity in it, moving it into a smaller platform. It won't go as far, as the battery has degraded to a certain extent, but they have a plan for actually extending the life and amortizing that battery back over a longer period with multiple platforms.

Now the consumer won't necessarily be able to do that, but there are other things that completely change the economics that we haven't touched yet, such as figuring out what the secondary use and market is. If we do have a lithium restraint we can recycle it and put it in the next battery pack. There are many different elements of this that will be game-changers that we can't see yet.

I am basing that on the typical technology trajectory and this is a technology we are talking about. I don't talk about this in terms of electric vehicles; I talk about it in terms of electric drive technology. I compare this to when logistics met the Internet, it is like these two things colliding -- and all of a sudden I can do something I couldn't do before. The expectation of the user is so dramatically impacted by it.

In 1995 I was having dinner with a guy from UPS who told me that they were spending $1 billion a year on information technology. OK, Brown (UPS) is a big company but I was really struck by that. I asked, "Is that because of the scale of the company?" He said, "Our studies demonstrate that 50% of the value to a consumer or a user of our service is information about where the package is." That was just mind-boggling to me; that justified the R&D and the investments that they were making in information technology.

CS: And the analogy to electric drive in the 21st Century is what, in your opinion?

BW: They drove certain technologies into the marketplace that would surprise you. UPS, Fedex, and so forth, are the guys that insisted that the barcode industry standardize the use of 802.11 wireless Ethernet. Why is that important? It used to be that for an 802.11 access point you would pay $4000 - 5000 and if you had to fit out an entire warehouse with these things, which UPS and Fedex did, it was quite expensive. However, because they needed to know where their packages were at all times, when it came off the truck or went to the warehouse, they needed to be able to retrieve items in real-time.

They didn't want to pay $4,000 for these access points, which they had to do when they were single-sourced. They drove 802.11 -- not just the standard, but the implementation of the standard through a group called the Wireless Ethernet Compatibility Association, which came up with you implement the 802.11 standard a certain way and you can put our sticker on it. Do you know what that sticker is called? WiFi.

The access point prices went off the cliff, WiFi got baked into everything, and suddenly we have a different world of information availability. But it was driven by wireless access through those guys. They insisted on interoperability.

Electric transportation is going to be very similar. I look at it from this perspective: In order to get off gas we have to have vehicle technologies moving in parallel with the fuel. The biggest problem over here is energy storage; in fact, that is the reason why gas won in the first place. It wasn't safety, it wasn't convenience, it wasn't that is smelled better; it was just simply that it was more energy dense.

Over here you've got the grid, now go back to my other statement that we are going to retire a lot of these coal plants because they are inefficient. Nobody owns the sun; nobody owns the wind; and how do we harness that in a grid of the future? It turns out that you run into the same problems here, because for intermittent technologies like solar and wind energy, you need energy storage.

So these two things are running on parallel tracks; they have the same problem and they start to converge. Electric vehicles are energy storage on four wheels. Once you go there it is like logistics meets the internet; you cannot go backwards. The macro economics drive this technology into the market place and you add to that the geopolitics, the environmental factors and behavior changing because the price of energy keeps stepping up.

The technology trends are obvious; we are going to continue to drive greater energy density and efficiency with these vehicles. Eventually electronic drive becomes your option, the question is how fast.

CS: And how much ecological damage are we going to wreak in the process? I always say that there are only a couple of questions. We are not going to be burning coal in 2050, and we are not going to be driving Hummers. Those aren't questions; the question is exactly how fast are we going to get there. How much ecologic damage will we have done in the process? And who is going to make a buck in the process? That's an open question as well.

On a different subject, I am interested in your perception of, for instance, zinc-air, lithium-air, and other battery chemistries. Can you speak to this?

BW: I don't understand the technologies; I am much more interested in the speed of commercialization. What is interesting to me is with all of these materials questions I look backwards and I see the work that was done for the materials that are currently being leveraged in anodes and cathodes in lithium batteries was done in Bell labs going back 25 years.

CS: It is a slow boat, for sure.

BW: So the question is lithium-air, the guy that runs Argonne National Labs -- he used to work at Bell Labs, and I would recommend you speak with him. He will tell you about lithium-air type advancements. Let's say that right now lithium-air batteries are a tenth of the energy density of gasoline. He thinks we can get half way across that spectrum, and that's huge.

CS: To be sure.

Last time I was in this city I saw you and Steven Chu at the same presentation, this was last spring perhaps.

BW: Yes. He helped us with the innovation motorcade.

CS: Right. He believes that we are fairly rapidly going to get to a price point and a level of energy density at which you have an affordable 350 mile range. He asked rhetorically, "Who wants to drive more than 350 miles a day anyway?"

BW: Again, if you step back and look at the whole picture, it is a dynamic one. Folks that are talking about the existing platforms, the Volt at $41,000, they are stepping back and saying how does that compare with a Chevy Cruise and why would I buy the Volt instead of that? I think that is a false comparison. The point is that if everybody who could afford one wanted to buy a Chevy Volt, we would have a hard time getting those out to market that fast.

That was the real problem last year. The drop-off in delivery points was July. That's because Hamtramck was changing over from the 2011 to the 2012 model and that's why they didn't hit their target. This year we will see how they do at that price point as they are rolling out nationwide. It's a fair bet that there will be some markets where this gets traction faster than other markets. To pick out one dealer, I am talking about Mike Kelly from Pennsylvania who happens to be a congressman, and says this is a ridiculous investment of tax-payers' dollars because nobody wants these cars: this is *not true*. They *do* want these cars but they want it at their price points.

I am delighted to be paying what I am paying for my electricity and I enjoy my ride. I traded in a car of equal value and I won't tell you what kind it was, but it was a high-end vehicle. People are going to buy cars for different reasons, but the things to pay attention to are what are the options for the consumer? How do you make certain that they understand the difference between a pure battery EV, a hybrid, a plug-in hybrid, etc., and make the right choice for their fuelling options?

As I heard Alan Mulally say when we were at the CES (Consumer Electronics Show) as a car maker and in any business the evaluation is a function of your discounted cash flow. You are going to end up basically cannibalizing some of your profit margins when you bring in a new technology. The question is whether *you* are going to cannibalize it or *someone else* is?

Not only are all of the OEMs pushing into this technology in a big way, but they are pushing in with multiple kinds of vehicles and multiple price points which are going to hit the market at a time when gas prices are going through the roof. This is what we are going to see in the coming year. You can talk to me in a year and see if I was completely out of my mind.

CS: Do you just see some set of international incidents spiking gas prices? Perhaps peak oil?

BW: No, it's pure economics. I am looking very tactically at the situation right now. The volatility of gas prices tells you that we really don't know what's going to happen next. The traders of this fungible commodity are just going to bat it around until what must happen is the global economy comes back online, that puts upward pressure. Then you look at stocks of oil and say what's our margin here? The answer is they are at historical lows.

CS: Speaking of refineries and oil companies, I had an interesting conversation with a friend who manages a hedge fund, and I say the long-term prospects that were are looking at ExxonMobil don't include the fact that they are going to reinvent themselves as a lithium ion battery separator company despite their public relations.

BW: But they own some of that technology.

CS: Yes but it is not going to move the needle in terms of long-term profitability. When oil goes away, so does that company, it seems to me. Is there a trajectory for a company like that to make good on its claim: we are not an oil company; we are an energy company? They *are* getting into natural gas I understand. But isn't it true, when and if your dream comes true in other words that we really have replaced oil, I would think that it would have to be bad for ExxonMobil?

BW: Do I care?

Electric drive technology is not necessarily going to be good for dealers, either. If you look at the sales and services model there is going to be a lot of disruption here. The point is we have built our entire transportation system on cheap gas, and gas is not cheap anymore.

And speaking of disruption: yesterday there was a bill passed in one of the houses in Washington State to charge electric vehicle owners (the pure electric guys) $100 because they are not paying road taxes. Interesting, but I am scratching my head, thinking hold on, we just adopted CAFE standards that are going to continue to exacerbate the problem that we have with the highway trust fund which is funded by gas taxes.

Let's make up our mind here. That's a discussion that will be happening above my pay grade, but I can tell you this:

with bridges falling down in this country and last week we had a major sign just drop out of the sky on Route 66. In Minneapolis, we have a crumbling infrastructure out there; we spent a lot of money in Iraq that we should have been spending on the road system.

Problems loom out there, meanwhile the price of gas goes up, the tax is the same per gallon; it hasn't changed since the early 80s and you don't want to say "taxes: in this town; you want to say "vehicle miles traveled" in this town, because that's progressive.

This is a battle that needs to be joined; we have two fundamentally opposing policies: one is CAFÉ standards that are encouraging efficiencies, and one that funds the roads out of gas taxes.

Literally we built the transportation system on cheap gas. If we are going to use less gas they are going to drive less, they have been doing that more and more because of the price of gas. They are moving closer where they work. These are all good things for electric transportation. The technology continues to advance, and it's a good fit for the 75% of people who travel less than 40 miles a day.

I need to come back to your question that these vehicles are so much more expensive, I disagree. The average price of a car sold in this country is $33,000. We are in range of that with a tax credit for many of these vehicles.

CS: It *is* getting better.

BW: So looking at it and saying I am going to compare apples to apples here and I understand that. But we will get closer in terms of that price premium, but it obscures the fact that many people can utilize this much lower, more stable fuel. It will be good for the economy, it will be good for the country and it will be good for the tax-payer to accelerate this process.

CS: There are those that say that the CAFE standards are a real threat to EVs, because what motivated you to buy an electric vehicle when you were getting 20 miles to the gallon evaporates when you are getting 60 miles to the gallon.

BW: I think the technology stands on its own. Again, when people get into the cars and drive them they want to buy them. That's the thing that makes me most optimistic. I start every interview with: "Have I told you how much I enjoy driving my car?" And I end every interview with: "Have I told you how much fun it is to drive these cars?" It's true. For me it's true; it is a totally genuine statement and that's what makes the difference for a consumer.

But the other thing is that we are doing a calculus based on a snapshot that is changing; this is a very dynamic picture. The price of batteries will get down to $250 per kilowatt-hour a lot faster than we think. Pike Research ... everybody's redoing their number all the time, because it is just not the cars; it's the grid also. There are tons of applications from buildings to cell towers that could utilize a more energy dense battery.

CS: Interesting. Thanks.

Here's a question you may not be to happy to answer, but let me ask it anyway. I am wondering about the motivation of the OEMs. I look at this thing from a macro perspective over a long period of time and there probably was some point at which some extremely senior people in both Big Auto and Big Oil saw their industries going separate ways. Big Auto probably looked at Oil and said, "We are not going down with the ship."

When I look at Carlos Ghosn, I see a guy who wants to be a leader in this space, and he is committed to it personally and professionally. But when I see other people in the auto industry, I don't know what to make of them.

You are right that every OEM that expects to be here by 2020 has an EV program at some level of development and implementation. Some of them obviously are further along than others, there must be some that are going along kicking and screaming. You are talking about replacement of an automobile that needs maintenance all the time with an EV that goes a half a million miles before it even needs a tune up. So dealers, at a bare minimum, must be resisting this transition in a big way, and they are extremely powerful.

Here's my question: You're the president of a membership organization whose members include the OEMs, and they rightfully expect you to represent their interests. So I can't expect you to say well these guys have dubious motives -- but isn't it obvious that some of these OEMs are not happy about this?

BW: I don't think so and I will tell you why. And first off I don't just represent OEMs. I represent utility companies, battery manufacturers, smart grid people, all the way up to the manufacturing chains to the lithium miners all the way down the community value chain to UPS, Hertz and Best Buy.

So I represent the community of interests. It's not an industry; it's a community that is coming together to electrify transportation. Of course we are partnered with everybody from AAA to Clean Cities; there are a huge number of interests that are motivated by different things.

CS: Yes, but one of those interests is the OEMs who say, "I wish it were the 1980s again because I want to be able to sell these people a new Buick every two or three years." For example, if you are a Vice President of anything at GM, you don't want this. Your 401(k) is rooted in your relationship with a customer in which they are more or less constantly buying stuff from you. Now all of a sudden you have a completely different calculus.

BW: Why is the OEM business any different than smart phones? At the end of the day, and this is the beauty of it, we live in an extremely competitive market place. The only thing that I would say is that I go back to my point before about vehicle technology and fuel technology/ sources need to be moving forward in parallel. If you go back to the whole conspiracy theories about automobiles and the oil industry, vehicles, infrastructure and fuel -- those things need to be moving forward in parallel. EDTA was founded by automobile companies working with utility companies; that is what got my attention. When you think about it, these are not two industries that work together very well. The truth of the matter is that we have to reduce all emissions from transportation and the generation of electricity, that's got to be done. These two industries can work together to bring this new technology to the market place. Neither the utility industry nor the auto industry is monolithic in its thinking inside of its own industry, let alone in my community. But what I can tell you is that where there is smoke there is fire. It's not just heat; it's light. These guys are truly trying to figure out where do we go in the future and these are massive challenges that we are facing. I don't think anyone in the auto industry would tell you that they think that the vehicle of the future is propelled by a combustion engine.

There may be debates over how do we get from here to there; there may be debates about at what point in that timeline hub motors become available and all kinds of other things.

I am now on the Board of a Group at the University of Michigan at Dearborn with some of the smartest systems guys in the automotive business. When these guys sit around the table, and they start talking about systems reliability and safety, it doesn't matter what company they work for; they are all smart people. They all care about who is driving that car and they all care about the infrastructure and how to get people safely from one place to another.

CS: I am impressed.

BW: It is extraordinary. Then I start talking about the power electronics which are about a third of the costs of a hybrid vehicle, or a battery electric vehicle, give or take; that's a *huge* thing and power electronics are heavy. What are they doing with the power electronics to lightweight them, to make them more efficient and faster? They do multiple generations between a generation of cars? The progress that those guys make with power electronics is astonishing. You can't turn that clock back, vehicles have been electrifying for decades; we just need to figure out the compressor.

And we *will* figure out the air-conditioning. There are milestones along the way; that just happens to be one of the key ones right now. Batteries will continue to advance, power electronics will continue to advance, the vehicles themselves and the systems will continue to electrify. I don't think anyone in the automotive business would deny that. I can show you dozens of PowerPoint presentations, and they all show that they are in a continuum to electrify.

CS: So when you said "compressor" you literally meant the compressor in the air conditioner?

BW: That's a sticking point right there, and for people that are much smarter than I am. That is a legacy of the combustion engine.

But the irony here is, and I don't know how a heat pump works, but recently we were talking about how to educate consumers in utilities. Again they are motivated by the concept that they have to create some smart load growth to pay for the investments that they need to make in the future in a cleaner grid. In a smart grid, in a more reliable and robust grid, they know they need to do that. It is like heat pumps, where they did a whole campaign around educating people to move in that direction.

CS: What a very interesting analogy. Well this has been fantastic, as I knew it would be.

BW: Well I am happy to be as opinionated about these things as you want. You are one of the few people I can get away with, being really opinionated.

CS: Ha!

Let me ask you this in closing. I have a colleague, a transportation visionary, one of these people who is constantly rethinking everything from urban planning, trams, mass transit, e-bikes and encouraging walking and biking. In fact, he was a guest on my last month's webinar; he was brilliant.

One slide from the webinar came from a question that I asked which was: Do we need to think of electric vehicles as a simple and direct replacement for the internal combustion engine? We take a four passenger car and we just simply replace the drive train? Couldn't there be more to this transition than this? Yes, this is the easiest way to do it because people do not like big changes.

You may know that when Edison replaced gas with electricity in the 1880s he was very careful not to introduce any more change than he had to; this is why he put lighting on the walls instead of the ceiling. He knew he could light a room a lot better from the ceiling, but gas lamps hadn't been on the ceiling, since they would have burned the place down. So he put his electric light sconces on the walls, where the gas lamps had been, because he didn't want to freak people out any more than he absolutely had to.

I presume with this paradigm shift there must be some analogy here, and that this is the reason we have a Nissan Leaf or a Ford Focus Electric as opposed to some more futuristic approach. Do you agree or not?

BW: Well before I answer the question I would just point out that we have the exact same thinking about gasoline and fast charging. Yes we are transitioning here and we need to look at the migration path, but I think what we are not seeing are certain technologies that are going to converge here and either take us in a slightly different direction -- or maybe take us back. Arguably, inductive charging in the road is like overhead electricity, with the side benefit that you are building a smaller lithium battery into a vehicle for the outlying transportation.

So in some ways it might be "back to the future" or "back to overheads." It's in the road. When people say, "Well that's far-fetched; why would we ever do that?" I say, "Go to Europe. They have the electricity overhead. You can build right in the road, and would probably be a lot cheaper in the long run.

I don't know what the accelerators are going to be. I know guys who say the amount of money that you can save just by leveling out the grid makes it possible for me to actually pay you for letting me slow down the charge to your car. Your utility might not want to do that, but the guys that are in charge of regulating the grid, the ISOs (Independent Systems Operators) might, and they might force the utilities to do it that way, because utilities are essentially run by regulators in some ways; they are policy driven.

There are so many elements of this that I think are hard to predict; it's like the electronics world, where you get convergence of various technologies when something becomes enabled, and it's adopted a lot quicker.

Pretty soon we are all going to be talking to our phones. And the interesting part about that is that in 1987-88 I saw an Apple video of a thing that looked remarkably like an iPad. A guy walked into his office, and he hit a button on the iPad and he started talking to it. That was Steve Jobs' vision from even before that. My point is that it takes that out-of-the-box thinking, and people are really inspired by that.

I am really shocked at the number of people who have stepped out to try and throw mud at electric transportation without coming up with a better solution for the challenges that we are up against. You can't tell me that we are going to continue to go forward in this country in this way.

My point is not whether we are going to electrify transportation, but how fast. Here is why fast matters. You have a 30-year mortgage; have you ever looked at the economics of a 30-year mortgage, and how they are front-end loaded? You pay interest mostly for the first couple of years, they have to show you what it costs you over 30 years that you have to pay, but what is the real cost of that money over that time? That's a pretty obscene number. Now if you put a little bit down in principal then you end up paying it off faster and the amount you pay is much smaller; this is a little bit like that.

CS: In other words the sooner you do something, the less it will cost?

BW: Exactly. And the costs that we are paying for gasoline – all the externalities all in, gasoline is at an unacceptably high price in this country. So I argue that if we can pull the deployment schedule in a little bit, it's like paying a third of your mortgage.

CS: Now I see. You are a man of analogies.

BW You are trying to save greenhouse gas; you are trying to displace petroleum, and that petroleum might be used for something else; I don't know; I don't care. What I know is we are going to, over those 30 years, need to continue to move hard goods and people that our economy depends on.

What is the American love affair with the car? Is it about freedom? Is it about access to opportunity?

The Chinese demonstrated this with electric drive technology in a very real way with electric bikes. They went from making a couple thousand electric bikes to making millions of electric bikes almost overnight; it was like within a three-year period. Electric bikes allowed people to go further and access economic opportunity -- to go to jobs they couldn't get to on a regular bike or by walking. They were quiet and they moved faster than a regular bike and people were getting run over by them.

So the municipal authorities actually started to ban them until they realized we can't turn back the clock on economic opportunity.

CS: How interesting.

BW: This is just a simple example of how people will find a way, and how they will embrace the technology that works for them. I come back to this: What are we doing here? The answer is we are trying to access those benefits and we are trying to pay less costs over those 30 years.

CS: Thanks very much, Brian. Profound stuff.

Dr. Jim Boyden—
Physicist

I'm both proud of and inspired by my relationship with Jim Boyden. Jim got his Ph.D. in physics from Cal Tech the same year I graduated from kindergarten. He went on to a brilliant and incredible varied career that included developing the laser printer business at Hewlett-Packard (making him arguably the most successful "intrapreneur" in history). More recently, and with greater relevance, was his work for Paul Allen's company Vulcan Capital, where he helped the team keep its finger on the pulse of clean energy technologies.

Craig Shields: Here is the premise of the book. If you were king of the world, you might prescribe one thing in terms of adding new forms of energy to the grid. In other words, you might say, "We need another 4,000 gigawatts of electricity worldwide, and here's how we're going to do it." Now, that may be a difficult question in and of itself, even if you had that liberty, but you don't. So, it becomes an even more difficult question and that is, "What will probably happen, vis-à-vis economic and political constraints?" Where do you see the possibilities, the imperatives, to make improvements in the way we generate and consume energy?

Jim Boyden: Well, I did an investigation on the subject of nuclear fission power. I will use the term "nuclear fission power" because people now know how to make that as opposed to "fusion power" which we do not yet know. I'm talking about the practicalities of fission power even though I'm not in love with it, because there are issues which people are quite aware of, especially with waste disposal of what is actually partially used fuel. I would modify our approach for favoring the development of nuclear systems. I won't just use the word "reactor," but nuclear power systems.

These are appealing to the developing countries as well as the fully industrialized ones because from the standpoint of many of the issues of energy concerning greenhouse gasses, emissions, national security, poverty and all these other issues. When they're compacted around the importance of having economically favored energy, we need to solve the problem in the developing world because if we don't, we will get overwhelmed. We in the industrialized world will get overwhelmed by the economic imperative of the developing world. They are not going to slow down development just because we're not supplying the right kind of technology.

Going back, I believe we should change our policies to favor the development of small- to medium-sized reactors and some of the peripheral issues around them such as waste management and some of these things. When I say "we," I mean the U.S. since that is about the only place you can have any influence. That's one of the first places I would go because I think there is a lot of need for nuclear power because of the lack of it. It can be done economically if you produce it on a factory basis; you don't want to have to restart the design every time you build one of these things. You can build them small enough so that you can transport them on a truck and rail in pieces. You can supply carbon free power in economical sizes to the world if you develop them.

One of the advantages of fission compared to fusion is that you can model it; you can predict how to build it. In some cases, there are some material issues, but mostly, you know how to build these things once you model them. It's a shame that we're not supporting that development, and favoring the commercialization of it and supporting it either through—I hate the word subsidies—but encouraging the development of these things.

CS: You're talking about uranium then.

JB: Uranium-based, yes. Classic-fission based things, but in smaller sizes. So, that's one thing.

CS: Well, in so far as it was your first answer, I presume it is at the top of mind. Most people reading this are going to say, "What about the safety issues?" We still don't know what's going on in Fukushima. We still have reason to believe this is a far worst disaster than anybody in Japan or anywhere else in the world is willing to admit.

JB: It's the problem of comparing reactors which could be built which are inherently physics-base safe. They cannot melt because the physics is designed into them so they cannot fail, as opposed to the stupidly built Fukushima one, and of course, the most famous Chernobyl. They were stupidly designed and built, and no one would ever do it again.

There are several designs floating around, but with no commercial incentive to develop them into the point of realization because of the extremely expensive deployment problems; people know how to build these things. We wouldn't put a Model T Ford on the road today because it doesn't pass all the NHSTA and all of the safety things and crash tests and everything else. It's like applying modern crash tests to vehicles from the 1930s.

CS: You're okay on the waste, i.e., the Yucca Mountain controversy? We still don't know where we're going to be sending all of this stuff.

JB: It's an important problem, but it has been vastly over-estimated. If you look at the number of cubic meters of waste that you have to handle and any decent deployment that would put a dent in the problem that we're talking about, it's relatively small amounts of stuff. It's important to have a solution for it, and there are a number of other things you can do. I won't go into some of the details of the research that's been done on how you compact this stuff and get rid of it.

Also, there are some efforts going on. One example that has been publicized and therefore I can talk about it is the thing that Bill Gates is investing in called TerraPower. You put this thing together, you bury it, and run it for 50 to 100 years and just leave it there. It's fueled once. It hasn't been proven yet, but there are a lot of models of them. There are some concepts which means the waste problem is minimized substantially compared to what we've been facing today. The problem is not to be minimized, but it's solvable.

I think I told you once before that I don't think we human beings came with a gene that allows us to evaluate relative risk. What is the risk involved with handling the waste of a nuclear reactor compared to burning billions of tons of coal and dumping stuff into the atmosphere—even so-called clean coal? We kill 50,000 to 100,000 people a year and don't even worry about it, because it's what we're used to. It's a relative risk problem.

CS: Yes, I see exactly what you're saying. But keep in mind also that people are concerned that there is a safety issue associated with *more* nuclear plants; you're talking about multiplying the number of them and not just the wattage. What about the issue of weapons grade uranium?

JB: This, again, is not an important problem. It has to be resolved by risk evaluation; by balancing risk.

CS: I see. What about thorium?

JB: I don't know an awful lot about thorium reactors. But we're not going to run out of uranium for some time to come. It's a different process and a different cycle. I am not an expert on what the balance is between those two things. It's a second order question compared to getting enough power within the next two to four decades.

CS: Let me ask you to put this in the perspective of renewable energy. The world is consuming 15 terawatts of energy as we speak, and we're receiving 6,000 times more energy from the sun than that.

When you add up not only direct solar in the form of PV and CSP, but all the subsidiary things like ocean thermal, tidal, ocean current, wave, run-of-river, and biomass. Then there are forms of clean energy that don't derive from the sun, like geothermal. I infer that you are saying that there really is no credible trajectory for the development of any of that stuff or you would not be talking about nuclear, is that correct?

JB: Oh, absolutely not. My answer is that I tend to concentrate on one thing. We should develop every single one of them in the order that makes the most economic sense. I thought I coined a phrase several years ago: "there is no silver bullet, so we need to fire silver buckshot." Then I heard somebody else use the phrase, so it's probably not mine. Anyway, that's basically what we need to do for the energy problem. There is definitely not a monochromatic solution, so we should do all of them.

The solutions are very regional. For instance, wave power is not going to work very well in Kansas. We need to use everything we can get. I believe in the potential of solar. We sort of glossed over the storage problem and keeping the grid alive. You know, all the little squares that people draw in the middle of Arizona in terms of the potential for distribution.

By the way, remember there are two or three articles that are important. One of them was in Scientific American about a year and a half ago where they claimed that we don't need anything else other than what we have now. The entire energy supply for the U.S. can be handled with solar in a distributive system. It's moderately credible, but an awful lot of things have to happen, both politically and geopolitically. There are people who claim the solutions are staring us in the face and we don't have to do anything further except expensive deployment.

CS: Right. I'm actually one of those people for what that's worth. As a friend of mine likes to say, "If you don't care what you spend for it, I'll get you all the clean energy you could use in a hundred lifetimes."

Let's get to the heart of the matter on this. To summarize where we are, you support an unnamed variety of different approaches, including nuclear.

JB: Absolutely.

CS: You're just saying we should do "x" if and only if there is solid economic trajectory for that. Are you entirely in favor of letting the market decide, or do you think the public sector should play a role in this?

JB: Counter to my general philosophy, which is fairly conservative I guess, I believe it's unfortunately true that we need, and should have, the government's involvement. This is one of the purposes of government. They need to make the decisions on things that, while they're not immediately commercializable, are still vital. The Internet is a classic example. That was developed for governmental, military purposes, and it became a huge commercial success. The same goes for the GPS satellite. Most of those things would not have happened, I believe, if there needed to be an immediate commercial need for them, if there had to be an economically successful operation.

I believe there's a strong reason for the presence of the government such as in the development of nuclear systems because of the huge uphill fight in getting nuclear deployed. The government should be playing a role in developing some of these inherently safe reactor systems, not to put a huge emphasis on nuclear.

The answer to the question about government is yes; it is a combination of the two. We can't sit around and wait for a venture capitalist to put $10 or $20 million into something that costs $2 billion to build. There's a huge gap. We're all waiting around for an enlightened high net worth individual to decide to deploy a substantial part of his or her fortune in doing this.

CS: That actually gets to the heart of the matter, which is following the money. We have to understand that large, powerful, economic forces are truly what's going to push us in whatever direction we go.

Maybe this means that we're headed in no direction at all. I run into cynics and pessimists who say we're simply going to pump every molecule of crude out of the ground and burn every lump of coal we can find until we've ruined this planet.

What do you perceive to be the major economic forces, and thus what is likely to happen here? You mentioned venture capital and you mentioned high wealth individuals.

By the way, the reason why I'm here is not only because you are a close friend and I trust and respect your viewpoints, but you are an individual investor on your own right. Moreover, you've played extremely important roles in organizations that (Microsoft co-founder and multi-billionaire) Paul Allen has run. In other words, I think that your credentials as someone who can answer questions like this are pretty strong.

JB: Thank you.

CS: Having said that, where do you see the role of angels, VCs, private equity, institutional investors, etc.?

JB: And the government.

CS: Yes, and the government.

JB: That's obviously a complicated question because there are many different roles that can be satisfied. To state the obvious, energy, as we all know, is the biggest industry in the world, mostly because of the fossil fuel industry. Therefore, making a dent and doing something positive requires inordinate amounts of money and you're struggling uphill against tremendous forces. There are protective forces that protect gigantic businesses. That attaches a cloud of reality over the whole issue of early-stage financing in terms of angel investors and so forth.

CS: Let me ask you to clarify. It seems you're saying, if I'm getting this right, is that the lobbyists who are hired by the fossil fuel industry create an environment in which angels might be thinking, "I'm not sure I want to play against these 700 people and the incredible influence they have in the outcome."

JB: Yes. Also, historically, the development of new energy related technologies and their deployment is a long cycle. The extreme comparison is with software or something where you can write code. I don't mean to minimize software developers, but you can write code and always do something new in the space of a few months or a year. But in the energy business, we're talking about ten years.

Most investors, except those with an inordinate amount of patience, are less likely to invest their money because they'd like to get a return much earlier, unless it's for partially philanthropic reasons like Paul Allen. He likes to invest in things that are going to have a reasonable return in terms of investment results, but he's willing to stir the pot and get things started just because it's the right thing to do. In some cases there is very little hope for getting a return. There are some people like this.

Angel investors usually treat it like a hobby. They're investing a relatively small portion of their net worth because it's a fun thing to do. It's not like an investment syndicate like venture capitalists. It's a single individual or a small number of single individuals investing in a new technology. That's really good, but most of the people are not interested in seeing through something that takes ten years to develop at the angel level.

Once in a while you're going to have such a stirring potential that people are willing to roll the dice and take chances on something like that, but it's hard.

CS: I understand what you're saying, and it does ring true in my experience as well. I don't know how much you're at liberty to talk about Paul Allen and what his past, present, and future plans are vis-à-vis energy-related investments, but I know readers would be thrilled with anything you can tell us.

JB: I have both legal and moral obligations to Paul since he has been very good to me; I have to handle this gingerly. He's made some investments in the area, some of which have not been disclosed publicly. He's not heavily invested in the energy domain. He considered a number of them when I worked for him and he decided it just wasn't for him. Paul does not like to substitute for the federal government. As I jokingly say, "They have even more money than he does."

CS: We're talking about a few tens of billions, right, in terms of order of magnitude.

JB: Right. His company deployed a lot of his money in the area of energy. He's one of the earliest people to recognize the importance of the problem before people started talking about climate change and all the geopolitical stresses that are all concentrated around energy. When I first started working for him directly in 2000, he was one of the earliest people who was really sensitive to how important this whole issue was. It hadn't really made primetime yet.

CS: Did you say he does, or does not like energy?

JB: He likes energy, but he's actually not widely invested in it because it's been very hard for him to find those things for which he felt—as he likes to put it—he can "move the needle." He's not interested in just making money necessarily because there are other places he can go to do that; he wants to do things that will trigger some other people to invest and allow *them* to make money. He's not heavily invested in the area by comparison, or what you might think, but he's been willing to take risks beyond what you might normally think of as a straightforward investment potential. I cannot talk about the extent of them because I don't know what's been made public.

CS: You would think of him as a "superized angel" then.

JB: That's probably close to it. He does have a venture group that assists him and guides the analysis of potential investments, but he steers the vote more strongly than the classic venture capital firm. It's not a case where people gather together on a Monday morning and a bunch of partners make a decision on what to invest in. They gather together and make a recommendation to Paul about where he should invest, but he's the final arbiter. After all, it's his money.

CS: I don't know the relative style or personality of any of these people, but the fact that it's one person versus a committee I would think would be one differentiating factor. Also, the fact of what are you really trying to achieve. If you're a VC, you want to hit a homerun, and you want to hit as many of them as you possibly can. You take as much technical risk off the table as early as you possibly can. It strikes me that these extremely high net worth philanthropists are probably more willing to take a little technical risk. In other words, they might say, "This may not work, but if it does, it will completely change the world."

JB: That's a good way to put it. Most venture firms are investing other people's money. They build funds in large banks and will raise a $200 million fund and then they allocate the funds in maybe a dozen investments or more. Paul's doesn't go seeking funds from other people; it's Paul's money. It's somewhat different than the classic VC.

It's not that VCs aren't rich; they may put their own money into it, but it's not their standard modes.

CS: Right.

I mentioned earlier what is commonly referred to as "Plan A." In other words, we don't have a plan; we continue with business as usual. "Plan B" is what you and I are talking about here, and that is essentially "grow our way out of it." Have you read *The Third Industrial Revolution* by Jeremy Rifkin? Did I tell you about this?

JB: I haven't read it.

CS: It's very good; this guy has been looking into this arena for a long time, and I find his ideas very credible. It's essentially the concept that the migration to cleantech can represent huge, sustained, planetary economic growth. I guess you could say that, at a micro level, this is what I'm trying to do, i.e., pulling together entrepreneurs and investors and saying, "Let's just get together and get this done."

By the way, there are people like Bill McKibben who think Plan B is total pie in the sky, i.e., Plan B isn't going to work and Plan A is going to leave us all dead. So, instead, what we really need to do is conserve and recognize that this is the end of the American Empire. We can't all live like kings.

Right now, you and I, as average adult Americans, are consuming 230,000 calories a day, more than 100 times the caloric intake of the food we're eating, with all the stuff we have: cars, electronics, industrial equipment, etc. This is just simply unsustainable for Americans, and God knows for the hundreds of millions of Asians who are becoming more consumer-oriented by the day. So people like McKibben point to "Plan C," which is biting the bullet and saying, "We have to figure out how we are going to enjoy our lives while consuming less energy." How would you respond to that?

JB: It's obvious what needs to get done, and part of our issue is how do you get people to accept this politically. As I have jokingly said, "Where is a really good benign dictator when you need one?" But what a benign dictator would do is totally unclear, because there are so many counteracting forces that people are trying to solve. The whole global problem is "don't make me do something I don't want to do."

CS: Right. In the United States that translates to, "I won't elect somebody who is going to suggest that there is pain in my future."

JB: Absolutely.

CS: Moreover, there are some who believe that there is essentially corruption. In other words, the lobbying process is so well-funded, so powerful and aggressive...

JB: That there is no hope.

CS: Right. Do you want to talk a little bit about politics and what you see?

JB: I am probably one of the most naive politicians you will ever run into, and it hasn't been one of the focuses of my life. Even though I do clearly recognize the importance of politics, I don't understand the system well enough to propose how you solve it. Let me throw in some of my pet peeves, which if there are enough of them, maybe they will give clues as to what might have to be done to solve it.

The Department of Energy is one of them. Many people, not just me, consider it totally ineffective, or worse, because they are spending huge amounts of money and not being effectively deployed. I had fond hope when Steve Chu was nominated and hired as the Secretary of Energy, but he's selfishly gotten ensnared in the politics. He is probably a reasonable politician, but whether he understood the depth of the problem, I don't know. To my knowledge, not much has changed which has been somewhat disappointing.

Since I was looking into the fusion sciences program, it disgusted me that there is roughly $250 million a year that is spent on fusion science. The way it was being spent was just ridiculous. The $250 million is spread over maybe 100 projects—I'm exaggerating. If I were King of Fusion, I'd try to figure out what the best four or five possibilities are rather than sticking your finger in the air and guessing.

Now, instead of supporting a university guy for $1 million a year, you're going to support somebody for $10-$20 million a year for five years. Then you may actually get a feeling for whether that particular approach might actually work. If it doesn't look like it's going to work, then we'll go onto something else, or if it looks like it shows promise then you start pouring more money onto it. If you distribute it widely to keep some professors and graduate students going, across a huge number of projects, you're not going to accomplish anything.

$250 million really could have been much, much better spent. We spent some of it in ITER (International Thermonuclear Experimental Reactor). Don't get me on my ITER soapbox. It's a $30 billion thing that is being built in Cadarache, France.

CS: Your take on which, I presume, isn't too glowing?

JB: It is a ridiculous thing. It is an experiment. It is not a power producing power plant; that will be another two to four decades. The results of ITER show that the first experiments aren't going to happen for at least two decades. I've lost track now, but it's going to be somewhere around a \$20-\$30 billion investment. They keep re-estimating the cost of the project, and it obviously keeps going up.

It's a deuterium/tritium fusion based thing. It's the "easiest" even though no one has ever done it. There are issues around D/T fusion. Tritium is a dangerous proliferation material. You add a few grams of tritium to a maverick nuclear weapon and it turns thermonuclear; that's not something you like to think about. People don't talk about it that much. They think D/T fusion is much better than "nuclear power" or "fusion power" because it's safe. Well, it has other issues. Even if it "works," that is you build a self-sustaining, burning plasma, it's many decades away from being a viable power plant. It turns everything inside radioactive.

Anyway, there are a number of issues around ITER. The only excuse that I can find for ITER is that if it does "work" and they build a self-sustaining fusion based reaction, it will open the flood gates for funding other fusion ideas that may actually be good. That is the only excuse I can find for it because at the moment it is acting as a giant sucking sound – where's the \$30 billion coming from?

The money is going to this one huge thing in France as opposed to spending $100 million each on another ten ideas. I think it's unfortunate.

CS: This is the $30 billion I hear is being spent on hot fusion. Obviously, there are low energy nuclear reaction people. What is your take on that, please?

JB: I'm not up to date on that. I did some investigation on that six or eight years ago. I'm not up to date on who is doing what. I think, as you know, it's almost like an underground science because people are almost embarrassed to work on something which actually might have the seeds to do something very special. I don't know enough about what's currently going on to give an intelligent or informed answer. I do find it very interesting.

I believe I mentioned that I think there are three questions to ask. Is there an effect—the so-called excess heat? Is the reaction understood so you can tune it and make it better? Last time I investigated it the answer was, "maybe there was an understanding of what is going on." The third question is: Is it going to produce enough power that it will make a dent in the key issues around energy needs for the world? That was my focus when I did my investigation. I was looking at gigawatts and stuff. My answer to the third question is "probably not." That is my intuition. That answer may not be correct.

If you produce enough low energy nuclear reactions, then maybe you would make a huge dent, but I've lost track. To my knowledge, the amount of money it takes for these kinds of experiments compared to what it takes to do the other kind of big energy sources that we were talking about is so, so trivial that some money ought to be put into it as long as it's properly vetted by so-called experts and other people who don't have a bias, but are going to look at it objectively. It takes such a small amount of money to do these lab-scale projects compared to what we are putting into big fusion and nuclear, that it deserves a chunk of that money to investigate because if it does work it is an important thing. I am highly supportive that some of the money should go into this, as long as it's properly monitored.

CS: As a matter of fact, Wally Rippel—the guy who I've mentioned a number of times to you; a fellow CalTech physicist—has told me that, and I think you've kind of alluded to it here, there is almost nothing more embarrassing for a scientist than to say that he's working on cold fusion.

Having said that, what Wally told me recently is that he wrote to Steven Chu a couple of times asking the DoE's position on the subject before he finally got a response. The letter that came back from one of Chu's staff members read: "We don't think this will make a difference." Wally goes, "Wait a second, that's not what you said originally. Originally you said this was a hoax; you said this was bad science. Now you're saying it won't make a difference? Those are two entirely different things. Why are you changing your mind? That's a red flag to me."

JB: Those are two very different things. That's sort of like my number three question. Is it big enough to make a big difference?

CS: By the way, you got your Ph.D. in physics at CalTech the same year I graduated from kindergarten. While I was learning to tie my shoes, you were publishing your dissertation — on what subject, by the way?

JB: What was then high-energy physics; making fundamental particles.

CS: Where do you see this going? If we were to have this conversation in 20 years, what do you think is the most likely thing we would be talking about regarding what happened over the previous 20 years vis-à-vis world energy?

JB: Well, 20 years, as you know, is a *really* long time. It's hard enough to project next year or two years from now. I think looking back 20 years people are going to realize that all those guys saying the pressures on the rate of production of oil were right. This is a serious problem. My God, the oil production is beginning to fall off worldwide. People are using more of it. The geopolitical tensions are likely to be very, very high around oil, and then water is going to be a big deal. It is now, but people just haven't risen to the surface, if you'll pardon the bad pun.

People are going to realize that the oil situation is really bad. There will probably be people looking back who say, "My God, I wish we would have done "X," like developing new reactors. What would have been a prudent thing to do 20 years before is now an emergency. I think people are going to be realizing that.

I think there will be some tapering off in the rate of use of energy. We're going to have cars that on average will get better gas mileage, but people have already shown statistically that when people drive cars with better fuel mileage, they drive more.

CS: There are certainly more of them; there's no way to stem the level of new drivers it seems to me.

JB: That's right, especially in India and China and those places. I think 20 years from now, some of the effects of global climate change and climate disruptions will, even though there is lots of noise in the system, begin to be realized. In fact, we will probably have improved our ability to do climate modeling by a substantial factor. Whether it's credible enough that people are going to make political decisions based on it is another question. People are very reluctant to make political decisions on scientific data if it's bad news. If it's good news then they say, "Oh good! You've told us the right thing to do." But if they're saying stop doing something then people won't believe the science and will point out all kinds of errors.

There will be a number of energy producing devices at a smaller level, but they won't be deployed in a substantial enough quantity to make a dent on the problem. What percentage I don't know, but I doubt if so-called renewables will ever really contribute more than a few percent or 10% or 20%. It's an 80% problem. I don't know what kind of new technologies there will be. I believe that there will be probably one or two fusion-based ideas that will begin to have some potential realization, but they will just barely begin to be deployed.

There will be a huge amount of attention to this problem and it will be driving some of the politicians' decisions. At the moment, of course, as an example, all of the politicians in our country and probably elsewhere are being driven by the immediate economic problems. Somebody mentions cap and trade and putting a 40-50% gas tax forward and so forth, and they receive no support for this. The other thing is we need to put everybody back to work.

CS: This has been terrific. Thank you. I knew this would be wonderful. Anything you'd like to say in summary?

JB: No, except to state the obvious and that is this is an extremely important problem. It may be the most important problem that faces the earth at the moment because of the multiple facets of energy like we talked about, including such things as population growth which is a non-linear effect, especially in the developing world. There is a strong correlation between fertility and poverty. Therefore, affluence is directly a function of the cost and availability of energy that is benign to the environment. All of these interlocking things make it probably the most important problem.

What is the old adage? Give me a hammer and everything looks like a nail. I've been working on energy for a long time, but nonetheless, these interacting elements are the essence of where we get the thing that drives our civilization and it will be crucial; yet, we're bouncing all over the place with respect to solutions.

CS: Thank you very much!

Tom Konrad, Ph.D. —
Alternative
Energy Stocks

Tom Konrad is another extremely intelligent individual I've known essentially since my entrance in the clean energy space. Tom's a financial analyst, portfolio manager, and freelance writer specializing in renewable energy and energy efficiency investing. He writes articles about investing in clean energy forAltEnergyStocks.com, and at the Green Stocks blog on Forbes.com. Tom's Ph.D. is in mathematics; his area of focus was "complex dynamics," a branch of chaos theory, which led to his conviction that knowing the limits of our ability to predict is much more important than predictions themselves, a lesson he applies to both climate science and the financial markets.

Craig Shields: We've had so many interesting talks through the years, that I thought I'd invite you to take readers through a few of your ideas associated with the migration to renewables vis-à-vis finance, mathematics, human psychology, and so forth.

Tom Konrad: Focusing mostly on the money side, because that's what this book is about?

CS: Yes, the whole thing is about "following the money," exactly. Maybe I can get you started. Here's what I believe; you can just tell me where I'm wrong. First of all, we're at the Renewable Energy Finance Forum, and I'm wondering if this conference isn't an exercise in "missing the forest for the trees." This is all about these extremely technical things like tax credits and master limited partnerships, things that are here today and gone tomorrow.

We don't seem to be looking at the bigger picture regarding where our society is with respect to clean energy and sustainability more generally. For instance, the way the utilities are regulated generally means that they have no interest in making a substantive change. The fact that we refuse to internalize the externalities of energy production shows we really don't care.

TK: Right.

CS: Those are the big issues that nobody is even talking about. Having said that, how do you get there from here in terms of politics seems to be a remarkably difficult question, especially considering that the forces that oppose renewable energy are so vigorous. They're spending tens of millions of dollars convincing people that global warming is a hoax, renewable energy is a job killer, and there is nothing the matter with fossil fuels, i.e, that there is nothing wrong with "business as usual."

TK: Right. I would say the forces opposing renewable energy are more basically the forces opposing change. These are of two types. There are vested interests—coal companies and oil companies who basically do not want to cut their business. Then there are individuals who want to keep on doing the same thing they've always been doing which, if you really wanted to move to a sustainable energy economy, part of it really requires lifestyle changes. They resent the sacrifices which would be required.

CS: I agree with that, but not everybody does. For instance (conference leader and president of the American Council on Renewable Energy) Denny McGinn told me earlier today the same thing that Jeremy Rifkin and Amory Lovins believe, i.e., that we can grow our way out of this. We just need to transition to this vibrant, new, clean economy.

TK: I believe that a vibrant, clean economy is a physical possibility in the sense that what they're saying is we could get most of the energy services we want with the available resources, or with a lot lower resources than what we currently use. However, I don't necessarily think that an economy can be achieved without physiological change and cultural change. I think, certainly in America, we have a culture of waste—a culture opposed to the economics of Rifkin. Rifkin is more of a techno-optimist, where I would say that Lovins is more of a systems optimist.

I don't agree that technology will save us. I do believe in the potential of efficiency and changing structures of how we do things. There are much more effective ways. I think 80% of all human effort is non-functional. It doesn't produce any positive result. I think that is one interpretation of Lovins' thesis, which is basically, okay, if 80% of our effort is wasted then let's just get rid of that 80%, and then we're only using 20% of the resources and we still got everything we want.

CS: Yes, I've listened to his presentation a couple of times in the last month or so. He has resource depletion curves that all go to zero by 2050.

TK: On the other hand, I don't think our culture is ready to do that. People don't like change, even beneficial change. As Lovins points out, there are all sorts of things we can do which are pure win-wins, for both individuals and society, but change is always scary. It's a visceral reaction which leads many to prefer the devil they know.

CS:Moreover, when I talk to people like this, I'm regarded as a socialist. As a matter of fact, it seems to me, and maybe this is unfair, but the whole right wing agenda to paint Obama as a Muslim food stamp socialist is also used against me when I talk about change and reduction of waste and addressing lifestyle issues in a more sustainable fashion. In other words, it is viewed as anti-capitalist.

TK: Right. And it doesn't have to be anti-capitalist. Much of what we need to do is improve markets. Improving markets is about as capitalistic as you can get.

First of all, an efficient energy market would actually achieve more efficient use of resources. That is, theoretically, what capitalism is about. The problem is that the way most Republicans seem to view capitalism really is a confusion between "capitalism" and "the status quo." This is wrong in many, many ways. First of all, markets in and of themselves are not efficient because humans are not rational actors. Second of all, we have all these structures that are built up over time that interfere with market efficiency such as companies doing regulatory capture.

There are structural reasons why we can't achieve this dream economy where we use so much less resources and have everything that we have now. The first reason I gave was the regulatory capture. Basically, the existing market structures have a life of their own in some ways, basically because the existing players want to keep the status quo.

Then there are also questions about investment. Do we have the right capital invested now to live in that world? In other words, if you want to replace a coal plant with a wind turbine then you have this big coal plant sitting there that you aren't going to be using anymore, and you have to invest in the wind turbine. If you want to make that transition rapidly, it requires a lot of capital which is in scarce supply right now. That leads you to the whole issue of guns versus butter.

On a societal level, many people are worried that if we recognize that climate change exists and it's a big problem right now then the only way to deal with it right now would be to go on a war footing and dedicate a lot of our resources to the transition, but we don't want to do that. It's easier to say, "Oh, climate change doesn't exist" than to say, "Oh no, this is a really big problem and we have to do a lot right now, but I'm not going to do it because I'm selfish." The human mind, instead, chooses to deny the problem.

CS: Yes, that's exactly right. That's exactly what Admiral McGinn told me just an hour ago. That's amazing. What about the fact that China is building a new coal-fired power plant at the rate of one per week, rendering somewhat moot what we do with our 700 power plants or whatever number there are.

TK: No, it's not moot. It's always compared to what? If China were building the same number of power plants and we weren't doing anything, we'd be worse off.

CS: Yes, you're right. It's not moot, but let's put it this way, changes that we make have a smaller effect given that they are, to some degree, dwarfed by the populations in Asia.

TK: What you're saying is that just because everyone else is doing it then we should.

CS: Well, no.

TK: But that seems to be the argument. Basically, I think that's really kind of an irrelevant point. It is a game theory problem, since you wanted some math. In some ways it is the prisoner's dilemma.

The classic version of the prisoner's dilemma is where two prisoners are accused of a crime, but the police don't have sufficient proof to gain a conviction. So they separate the prisoners, and offer each a deal. They tell each, "If you admit guilt, and your accomplice does not, we'll give you a 1 year sentence, but he'll be put away for ten years. If you both admit guilt, you'll each get three years."

The cops have set it up so that each prisoner is taking a big risk if they do not confess, at least if they don't trust the other prisoner. So they often get what they want: two confessions, when all the prisoners had to do to get away entirely was to keep their mouths closed.

We're in the same situation with China over greenhouse gas emissions. Everyone would be better off if we all cut our carbon emissions, but if everybody cuts their carbon emissions except for one person, then the person who doesn't gets the most gain.

So, how do you solve the prisoner's dilemma? Well, you solve it by repeated games of prisoner's dilemma. In repeated games of the prisoner's dilemma, one of the prisoners has to make the first move. We have to stop ratting the other one out so he can learn to trust us, and we can move to the optimal outcome of everyone free from global warming.

If we make the first move, and start cutting our own emissions, there is no guarantee that China will follow, but if we don't make the first move, why should they?. And we're not even the ones who have to make the first move... the first move was made by the signatories of the Kyoto protocol. Right now, both the U.S. and China are the bad actors making sure everyone does Global Warming prison time. Sure, it would be nice if China stepped up to the plate before we did, but it's our responsibility to make it easier for them by taking some responsibility ourselves.

We are the only global superpower. It's time we stopped pointing fingers and started acting like adults.

CS: Is that a good analogy, in the 21st century, with China and the U.S.?

TK: I think one of the problems is that each player is not really the same person every time they come to the table. The U.S. is a different person every two to four years depending on who the political people are, and that lack of continuity leads to a lack of responsibility. There are very few watchdog countries to their word, either.

But I think if you say just because everyone is doing it then I'm going to do it too then you're making excuses. Somebody has to be the first person to break the cycle. Europe took that step; it's time for us to join them.

CS: You're right. I think, culturally, we in the United States are nowhere near the Europeans, for instance, with respect to a sense of community.

TK: Yes, seeing commonality. We have this individualist culture, and I think it is causing problems now. I think that is a problem, and I really don't have the solution for it except for trying to change the debate and make people realize that, yes, we should be taking the high ground. If we want to see ourselves as moral leaders in the world, which the U.S. used to want to do, that's what we have to do. We seem to be yielding the moral high ground and taking on a much more "me" mentality.

CS: I certainly sense that. Do you think if we polled 1,000 randomly chosen Americans who are at least 45 years old, do you think we would find that?

TK: I think 950 of those 1,000 random Americans would all say that, "Oh no, I always take the moral high ground," and 50 of them actually do. But they would not be lying; they would believe it.

CS: Well, that's my point. It's not a matter of admitting the truth; it's a matter of recognizing the truth.

TK: We're lying to ourselves, and we do that because it's often easier than admitting the truth. Our current media allows us to do that.

CS: Dennis McGinn has been around legislation that came initially from scientists studying seatbelts in the 50s, DDT in the '60s, and asbestos in the '70s. He says that you can replace the names and decades and the rest of it remains the exact same thing. In other words, the whole process by which people say we have a problem, industry denies it and comes up with extremely well decorated and financed debate structures around why a) the science is clumsy, and b) even if the science is right, it will destroy our industry and it will put people out of work if they're run any differently.

TK: I think industry is the problem, and that is part of what I see as my mission: to be someone who helps people invest in those parts of the industry that aren't the problem. Industry is owned by investors. Until we start putting our money aligned with our morals, our industry won't be aligned with our morals.

That's not enough; there are agency issues as well. The CEO who is running the industry does not always act in the best interests of the shareholders which is another common market problem that has nothing to do with clean energy or anything. Just because the market is imperfect doesn't mean we shouldn't be trying to push it into the right direction with our investments.

CS: Are you talking about socially responsible investing (SRI)?

TK: In the broader sense, yes. The problem with SRI is that we all have different morals and so it's very hard to define. I think, simply, what you need to do as an individual is you need to put your money in things that don't contradict your own beliefs. You can find an SRI fund that generally lines up with your morals and that may be a good way to do it. It depends on how much time and effort you're willing to put into what you're investing in.

This is another case where we are denying our call to act because it's too much work. Investing is work. A lot of people are afraid of money and so they abdicate their duty to invest morally by buying the line that markets are efficient, and if you simply invest and make the most money then you are doing good. That is part of the market orthodoxy that has been disproved so many times it's not even funny, but still, it pervades a lot of thinking about investing.

CS: Yes, I'm sure that's true.

TK: Again, it's very similar to climate change. We don't want to do the work and therefore we will believe whatever allows us not to have to.

CS: Let me ask you this, and this is something that I'm going to try to pull apart in some of the remaining interviews that I have for this book. It strikes me that a lot of the problem of integrated renewables in the creative mix is the way the utilities are regulated. In other words, they simply don't have an incentive to do this.

TK: Yes, however, it is very hard to speak in general terms because there are more than 50 regulators. There are 50 regulators for the publically owned utilities, but then there are also municipal utilities and co-ops. Many of them do not have the right incentives.

CS: Let's take an example. IOUs are the largest in terms of terawatt hours on the grid in the United States, so let's start with them. Can you tell me about the 50 regulators and how this works?

TK: I can tell you a lot about Colorado because I actually testified before the Colorado PUC a few times. They were in the process of transitioning to try to give their utility incentive to do what they wanted to do. In the past, which I think is a very typical model of regulation, the utility was mandated to provide everyone who wanted it, power. To make sure the power doesn't go off, and do it at the least possible cost. The definition of "least possible cost" did not include any externalities. Furthermore, if they did that then they were guaranteed to regain all of their capital at a certain rate, and it was based on a per-kilowatt-hour sold. So, if they sold more, they would get more money.

Now, on top of that, there was some regulatory capture. When you're in the resource development process, it all starts off with the utility modeling how much power will be needed, and figuring out what the cost of everything is. They put in their own numbers. It's based on reality somewhat, but they have an incentive to spend things.

CS: This is the part I want to understand. It's return on equity in other words. They invest a billion dollars in a plant and they can amortize that; they can pass along those costs.

TK: Yes, and they can also separately pass fuel costs directly through to consumers. If there was any risk in fluctuating fuel costs, they didn't have to worry about them; this is sort of the old regulatory model. They generally would have projections. They come up with projections of what fuel costs are going to be and what the cost of building the plant is going to be and how much power they're going to need. Those projections are always wrong, and they also require judgment. They are probably the most knowledgeable group at the table. Once they come out with this plan, everyone else comes in and they can nit-pick. "Everyone else" includes big consumers like steel plants, the Office of Consumer Council, etc.

CS: What about advocates for the individual consumer?

TK: Yes, there is a state consumer advocate, but their mandate is mainly low cost. The consumer advocate also tries to make sure that everyone is served equally; they're not biased. Then there are the advisors to the public utility commission—technocrats who will also give their opinions. They're trying to figure out exactly what the utility and other parties are saying and what they are trying to spin, but they're not nearly as well funded as the utility itself. By the way, all the utilities' costs for these projections are paid for by the rate payers; yet, they have the best funding. It's really the rate payers paying money to put up this projection which may or may not happen.

Actually, in Colorado—and this is fairly unusual—any individual rate-payer or group can also testify. Again, any funds for that have to come out of their own pockets. You have one group, the utility, that has a lot of interest in this, and then you have all these other groups. You have the bureaucratic groups who are supposed to represent the people, but their mandates are very strict. Then you have individuals who can come, but of course, their interest in it is not to the level of the utility because they're just small players. The large consumers do usually have enough funding to really get their voice heard.

There are a lot of conflicting interests. As you can see, there's really no advocacy there for renewables, especially historically, when renewables were higher cost. Now, what changed over the last half a decade was that the Governor of Colorado who appoints the PUC decided we should include externalities. He wanted to include their externalities, and said they would go through a small rulemaking process to try to actually include these in the decision-making process. We're no longer going to go for least cost; I forget exactly how he phrased it but various externalities were included. This was contentious, by the way; there are people who argued 'Oh, these externalities are zero or they are very high.'

It ended up with the externalities as a "fixed adder." I'm thinking about an energy efficiency case where they are trying to determine the externalities of a kilowatt-hour and they decided on five or fifteen percent of the cost; we're just going to increase the cost by five or fifteen percent and that will be an adder—very low but at least it's in there.

CS: I see. Now what about intermittency and the cost of integrating a power source that has variability issues.

TK: At low penetrations it is irrelevant because the utility already deals with fluctuating demand. Once you start getting beyond fifteen percent for wind it starts becoming an issue. Initially you can deal with that by averaging out. If you take wind over a lot of area, combine it with solar, it's much smoother than it is if you just look at a single point. You can get up to maybe 20% with existing protocols to do that.

And we're fairly close to this now in places like Colorado, which has a lot of wind. They are heading for 20%, and I think they are 15% now. Once you start getting past that 15% barrier you have to start changing protocols. You have to integrate in weather forecasts so that you know, yes, wind is variable but it's a lot worse if you don't know when it's going to change. You have to have better systems for monitoring the wind, monitoring what is happening at this turbine because it will tell you what's happening at the downwind turbine.

Then you have to start looking at control areas. Utilities often have mandates that over a small area everything has to balance out. If you *broaden* that area, you have more flexibility. Actually that's just a bureaucratic thing. If you broaden these control areas, that's another thing that you can do to allow more renewables on.

After you get past 35%, that's when you probably need to start looking at storage or you actually need to start building a lot more transmission lines so that you can broaden your area even more.

CS: How great a challenge do you believe this represents? Obviously we are a million miles from 35% in any state in the union, but for instance when I look at electric transportation I often say, 'Well, I don't see the problem here. I see three things coming together nicely. I see the availability of EV's from OEM's, the build-out of the grid, and consumer acceptance all moving along more-or-less in lockstep; I don't see any one being terribly out of phase with the other two. I bring that up to ask you if you see that analogy here in integration of large scale renewables. Do you see a natural build-out of the grid occurring simultaneously?

TK: No, I think localized problems are inevitable and they are already happening. I think overall it's not a big deal but last spring, a year ago, the Bonneville Power Adminstration had to curtail production of wind because at the same time there was a lot of wind, there was a lot of hydro because of melting. There was just too much power and they didn't have the transmission outside of Washington State to get it where it was useful.

CS: Isn't that fairly common, i.e., curtailing a significant amount of wind every night?

TK: No, it's unusual that they don't produce when it's blowing. Because you're at low penetration you can tamp back your coal plant; but you can only do that so much and it's not good for the coal plant. You can certainly tamp down your gas plants. There is a lot of natural variability in the system.

In this particular case in the Northwest, they couldn't tamp down the hydro because of environmental laws where it could cause problems for fish. They wanted to tamp down the wind, and the wind producers were very unhappy about this partly because of a poor incentive for wind, i.e., the production tax credit (PTC). Wind producers are often willing to be paid a negative price because they get 1.2 cents from the federal government. If they have to pay half a cent to put power on the grid, that's OK with them.

But it's not displacing coal at that point. The wind producers there wanted to displace hydro. Where is the environmental benefit of that? There isn't one and so we're subsidizing something that is nonproductive.

CS: I completely agree with that.

TK: And that's a problem with the subsidy. Part of the reason all those wind turbines were built there was that was the windiest spot. They were all built in one little valley and guess what: the winds in those valleys tend to be strongest in the spring and coincide with the snowmelt and peak hydropower production. Well, there are other parts of that region where wind is not as correlated with hydro production and at those times power is more valuable. Because we chose to use the PTC to subsidize energy production, of course the wind developers go exactly where they can do the most energy production but not the most useful power.

CS: Well it does go back to something that I've noticed and that is, I believe in subsidies when it can be demonstrated that they are for a clear, unarguable social good. To take the obvious example, we're still spending tens of billions of dollars of taxpayer's net worth to make the oil companies richer. This is a 90-year-old industry that is extremely profitable and it's stable beyond measure. That's the obvious one but there are less obvious ones and this current example is a good one.

TK: I think it comes back to the fact that fiddling with markets is trouble. It causes trouble and then it is very hard to unfiddle with them. I think the PTC is better than no PTC but it's clearly not a perfect market position.

CS: But let me ask you this: how hard would it have been to say, 'Look. The goal here is not to build wind for wind's sake. The goal here is to displace carbon, and we're not spending a nickel on anything that doesn't do that.'

TK: What about displacing nuclear? Ideally, the goal here is to get the energy services we want with the least environmental impact.

CS: Yes. Well, I guess what I should have said was we're not spending a nickel if it can't be proven that it's a social good. In other words, if it's not displacing ...

TK: That's a problem too. That goes back to what I was talking about. There are a lot of things that we know are social goods that we can't prove are social goods.

CS: Give me an example.

TK: For instance, there is a whole argument about killing birds. Let's suppose the social good is not killing birds. We can prove that wind turbines kill birds because we find dead birds at the bottom of wind turbines. We also are pretty certain that coal plants kill birds because of pollution and climate change which will destroy the entire species. But it is really much harder to measure. If we only look at what we can prove, we sort of come back to this least common denominator.

CS: I guess my standard of proof would be a little bit more liberal than that.

TK: Yes, but once you put it in legislation the lawyers are going to go after it. Yes, 'I know it when I see it' is a really nice criterion and that's what I use when I'm looking at investments but the only thing I have at stake is my own money. Once you start legislating you can't use that.

CS: Right. I'm reminded of the difference between the rules for deciding civil cases versus criminal cases. Criminal cases require proof beyond a reasonable doubt; civil cases are decided on the preponderance of the evidence, which I find pretty compelling in cases like the environment. For example, I often ask people, "How much evidence of climate change do you need to see?" But that's just me.

TK: Well, yes; again, that is just you. It is you and there are a lot of other people who find the evidence against climate change very compelling.

CS: I guess you have a point.

TK: Back to the subsidy issue, there are a lot of things we can do that both improve markets and improve environmental good. Those are the things we should be concentrating on now, though I don't think this will solve all the problems. Amory Lovins' has a point with his thesis too, that efficient markets will solve all the problems. I disagree with him that it will solve all, but there are a lot of things that we can do that will improve. Increasing information is also critical, for instance; whenever a piece of real estate is sold, good energy data about that real estate should be given to the buyer. That clearly improves markets because it improves the information of both participants. Lack of information is one of the biggest barriers to functional markets. You could mandate better information; it could reside in a national database; there are all sorts of ways to get better information. These sorts of things will improve markets and they will improve environmental performance because once I know the energy performance of my house is going to be seen by my buyer, I have a much stronger incentive to improve it.

CS: Yes. Even if you're not selling your house, once you know what those data are, that's a good thing; it's like measuring anything else.

TK: That's another aspect, yes. If we don't measure it then we won't improve it. I do think there are a ton of things we can do about information that would greatly improve markets. I think Opower is a great example; it's a company that sells a solution to utilities that gives customers a little information on their bill about how they're consuming energy relative to their neighbors.

CS: Yes, I've seen that.

TK: They give you a little smiley face if you have low consumption and a little frowny face if you have high consumption.

CS: But it's opposed on privacy grounds.

TK: Yes, better information is often opposed on privacy counts, but it's really opposed by people who have an interest in a dysfunctional market. I believe that it's probably liberals because liberals don't like functional markets. A true conservative likes functional markets and therefore they would support Opower's happy faces and smiley faces. It's those damn Socialists.

CS: Ha! When I think of liberals, I think of do-gooders -- not in the pejorative sense, but people who concern themselves at least in part with the well-being of other people outside of their own family, outside of their state, outside of their nation. So I would think that, in that sense, a liberal would support anything that created a better world for everyone.

TK: A liberal will support anything that is clearly (to them) creates a better world for everyone. But, in my experience, they tend to focus too much on immediate gain and not enough on global gain. If they see a person hurting, they say 'No, we don't want to do this policy.' That's often not in the best interest of society.

CS: Well, I can certainly see that the concept of sacrificing the good of some for the benefit of a greater number of people is a reasonable idea.

TK: Right. This is why I'm willing to kill birds with wind turbines, which we can minimize, in order to kill less birds with coal plants.

CS: Exactly. Moreover, a bird is 800 times more likely to fly into a plate glass window and 1500 times more likely to get killed by a cat than to be whacked by a wind turbine.

TK: Again, we're focusing on the obvious, immediate pain and that can often be a barrier to community good.

CS: Yes. What else have you been thinking since the last time we got together with respect to this whole subject? Let's put renewable energy as a subsidiary subject to that of sustainability.

We're at a finance conference, so maybe this should be astonishing, but I think you could count the number of times you heard the word "sustainability" in this two-day event on the fingers of one hand. We heard a lot about national security and the most obvious imperatives but I think maybe we choose our battles carefully and we choose our words to support those battles. Sustainability may not have the cachet here that it would have someplace else. After all, again, this is a finance forum.

But what scares me most about the way our culture is headed is that sustainability is written off as a fad, as today's buzzword. I think that it's the most important concept of the 21st century.

TK: I agree with everything you just said. I think a large part of the reason that in a finance forum we are not talking about sustainability, which I consider to be a moral issue, is because we don't have a moral investing culture. As I said earlier. I basically see that as a moral failing of most people who are involved in investing.

CS: Not to overly get philosophic about this, but I guess it goes back to what we were just saying about liberals. In other words, in addition to talking about other people who are living now, we are also talking about the as-yet unborn and the recognition of our responsibilities to leave them the same quality of life that we had.

What else have you been thinking with respect to the migration to renewables and to a more sustainable future?

TK: I think we're going to get there. I see there is a lot of hope that we will get a long way without the structural changes that I see would be necessary to get all the way there. It inspires hope seeing just how quickly prices of renewables have gone down and I think that there is some agreement on improving efficiency. I also think I see some hope in terms of better market information.

CS: In terms of efficiency, people often say that this is the "low-hanging fruit." It's an area of great potential environment improvement, and you put a lot of people to work.

TK: Yes. I think that the net gain of green jobs from solar actually is questionable, but I don't think it's at all questionable when it comes to jobs from efficiency. The reason it is not questionable is because you are saving money and your efficiency is correcting a market failure. You pay less for the energy services you have after you've taken efficient resources. When you pay less, you have more money to spend on other things, and that money creates jobs. There are also jobs created when you do the actual spending; any actual spending will create jobs.

By fixing the market failure that exists, by applying efficient solutions, you have created a new economic surplus. That surplus adds to the net well-being of society. That will create jobs.

CS: It raises a question about job creation generally, though, it seems to me. Take the Bush tax cuts for the wealthiest Americans; this is normally applauded because of jobs. In other words, we still believe in trickle-down economics, in which wealthy people hire other people. But I don't find any kind of relationship; you don't hire people that you don't have to hire simply because you have more money. You may have a house in Bermuda but you certainly don't hire people that you don't need to hire.

TK: I believe that if you make the market more efficient you will create jobs. Taxes are, in general, a drag on the economy because the government is bad at spending money. Of course, business is bad at spending money too. It's hard to say; there isn't strong evidence either way on that.

CS: Well, going back to jobs. You're saying that the one unequivocal winner, vis-à-vis this transition, is efficiency and its effect on the number and quality of jobs. Both from the standpoint of the obvious, immediate putting people to work and the multipliers.

TK: Right. Conventional economics tells you efficiency is a winner. On top of that you get a bunch of externality benefits. Now the other renewables may or not be winners (in terms of job creation.) That depends on what you think is important. Since all energy is distorted, it's very hard to say what's best. It depends on your goals, i.e., which externalities you think are most important.

I generally favor reducing carbon emissions. I'm generally in favor of reducing traditional criteria, i.e., pollutants which lower well-being. Again, all those things you have to put values to; it's very hard; I feel you can only do it qualitatively. Once you try to do any of this with precision you're basically going to be very precise but you're also going to be precisely wrong.

CS: In other words, there are unknowables that are significant.

TK: Yes. I think the unknowables are larger than the knowables. But we know which direction they're pointing in. If we wait until we know everything with certainty before we act, we'll end up knowing with certainty that we acted too late.

CS: Well, this goes back to my thing about trying to hit the broad side of a barn. You don't have to have too much quantitative analysis to understand that coal is a bad thing. There is the Harvard Medical school report about the $700 billion associated with the health and environmental costs; how accurate that number is remains to be seen, but it's very large by any measure.

TK: I'm sure there are places where using coal produces enough societal good and there certainly have been times that it made sense to burn. Right now, I agree with you. It's fairly clear that we're overdoing it. Again, it's just very broad. At one point it made sense.

CS: Well, it certainly made sense before we realized what was going on vis-à-vis sustainability and certainly climate change.

TK: And we were also doing it on such a small scale.

CS: DDT made sense to me in 1961 when there were mosquitoes all over South Jersey and we wanted them gone.

TK: And it still makes sense to sell DDT soaked mosquito nets in Africa.

CS: Simply because the malaria is such a high cost.

TK: Yes. A lot of these debates, I really think, are based on people believing what they want to believe. A lot of things are arguable. I think Democrats often focus too much on individuals and individual pain while Republicans focus way too much on rich individual pain. In truth, we really don't want *either* sort of pain.

I really try to focus on things that are win-wins. There are a lot of places where there are not. Again, I actually think renewable energy is like second tier in terms of what are win-wins.

CS: Efficiency, you mean?

TK: Efficiency and conservation. I think changing the structure of our economy is better. Renewable energy requires inputs. If you look at energy return on our investment, you have to put in between one unit of energy for every five units you get out; it's 1.2 if you're talking about ethanol.

If we want to operate our society at the same level we have been, we need to be able to put in one unit of energy for every 20 or 30 we give out. Renewables can't do that. Basically our society was built on free energy because we were taking stored fossil energy. What can do that is better use of those energy services, like the example I gave you of improving market mechanisms. Often you don't have to put any energy in and you get more services. That's what is going to keep us at a higher operating level.

I think one really rough way to think of energy returns on our investments, how much energy you put in and how much you get out, is if your society is operating at a one to five energy return on investment; one in every five units of energy in your economy is going back to producing more energy. That's 20 percent of your economy devoted to producing energy which is nonproductive. It doesn't help the human welfare; it's a drag. As you go lower down that chain it gets worse.

And it's a lower benefit for humans. We're used to living at one in 30—only about three percent of our economy is, but that's gone up. What we have to do is get a lot more benefit out of the four units that are left.

CS: That's interesting. Where is this going, with respect to fossil fuel? We talk about the Keystone XL pipeline, for example. I'm told this is extremely expensive, not only financially, but ecologically to get out of the ground, to move, to process, etc. Where is that with respect to that one to five ratio, if you happen to know?

TK: It's pretty low. I don't know but it's definitely below 10 in terms of how much energy you put in. It will never be close to the one in 30 that we used to have with oil. And it's pretty dirty.

But I personally think that the whole Keystone debate was really a waste of political capital, and one I consider a side issue. I think that oil up in Canada will build a pipeline to their West Coast; the oil will go out one way or another. We sacrificed a lot of political capital stopping it. I don't think it really made much difference; it delayed it a couple years. I would rather have seen that political capital put to something more useful. I think the environmental community was just so desperate for a victory, any victory, that they saw this as one potential victory. I think it really was a "win the battle, not the war" type thing.

CS: It's interesting. For instance, it's common knowledge, whether it's true or not, that Obama has spent so much of his political capital on healthcare that it's damaged him in terms of being able to get done some of the other things that he may have wanted to or maybe should have done. You're saying there is something of an analogy here. In other words, fighting this battle weakened the environmentalists. Is that your position?

TK: Well, I think whenever you make a cause so important, you have to trade political favors for things. Is that something we really wanted to trade political favors for? On the other hand it may have gotten some people jazzed up and helped political support. I don't know. Having a victory is important, but I don't think this in particular was that important. I think there could have been victories that were more crucial.

CS: In closing, how would you summarize the state of the industry? People wonder if these clean energy projects are drying up, and actually, there are challenges associated with the partisan politics, for sure, which means the end of congressional support for various programs. There are some black clouds in the sky but solar installations are up 109 percent in the last 12 months.

I don't think this is as dark as some people think, do you?

TK: Well, it's much cheaper than ever before, which has driven some companies out of business. I think that's what people see. I think people often conflate the fate of individual companies with the fate of an industry. Often those go in opposite directions. The fact that prices have come down so much is great for the solar industry, but it's horrible for the individual companies.

Although it's great for new emerging manufacturers that can survive the lower prices. It's the best of times; it's the worst of times.

CS: Great, Tom. I really appreciate it.

TK: Happy to help.

Jerry Taylor—
CATO Institute

Jerry Taylor is a Senior Fellow at the CATO Institute, and among the most widely cited and influential critics of federal energy and environmental policy in the nation. Taylor is a frequent contributor to the *Wall Street Journal* and *National Review,* and appears regularly on CNBC, NPR, Bloomberg Radio, the BBC, and Fox News. His op-eds on public policy have appeared in the pages of *The Washington Post, The New York Times, The Los Angeles Times* and most other major dailies.

I found this to be the most interesting interview of them all, precisely because, as the reader will soon see, Jerry and I don't see the world in even remotely similar ways.

Craig Shields: Jerry, thanks for taking time with me here. I have listened to a number of your recent talks to familiarize myself with your work. The other day, I wrote on my blog something like: "This guy is brilliant, and he is an unbelievable presenter. Now, it's true that we don't agree on much, but I do think there is legitimacy to a lot of his ideas."

As you know, the current project is "Renewable Energy— Following the Money." It's an attempt to answer the question: If it's true that there are a great number of imperatives to move away from fossil fuels, then why are we doing this so slowly? Why isn't this happening faster?

People have explained it in a number of different ways, and I figured that you would take a completely different tack and assert that the so-called "imperatives" are not what I think they are, that the costs are far greater than I recognize them to be, and so forth. I don't want to put words in your mouth, but let me begin by asking you to articulate why you are not a big proponent of renewable energy.

Jerry Taylor: If renewable energy makes economic sense then capitalists will invest their own resources into it and the government need not do anything. Profits are gained by making smart investments in emerging technologies, and the reason we don't see that much investment in renewable energy absent government incentives is that it's light years from being competitive with natural gas in the electricity sector, and gasoline in the transportation section.

CS: But coal is the cheapest of them all.

JT: I'm not sure that is really true. At the margins coal is irrelevant; nobody is building new coal-fired plants. 95% of generating capacity built last year was gas-fired and that has been the case for the last couple of decades. Investors in renewable energy are competing almost entirely against natural gas, not coal, for new orders. Coal survives because of very complicated regulatory incentives to provide extra life to old coal-fired powered plants, and that is a whole different story.

So that is why we are not seeing the market move as quickly towards renewable energy as proponents would wish: the economics just aren't there. As far as your first point, there is an imperative for various reasons to move away from fossil fuels, I take issue with that; I am not entirely sure if there is any such imperative. From an economic perspective if it is the case that there are uninternalized externalities associated with fossil fuel consumption, and there seem to be some, then the correct remedy is to internalize those externalities through the price mechanisms.

I think 100 out of 100 economists you talk to whether liberal, conservative, Democrat, Republican, environmentalist, anti-environmentalist will agree with that proposition.

CS: That would be more than fine with me, not that I count myself as an economist.

JT: The best remedy is to get the prices for fossils fuels correct and then let the market sort it out. The environmentalist assumption is that the externalities, primarily the environmental externalities associated with fossil fuel, are very large. But if you look through the literature you find that those who try to estimate those externalities find estimates all over the place. There is no consensus as to what the externalities might be, and the reason for that is that there is a great deal of uncertainty in the scientific and health arena about what the impact is of incremental increases in fossil fuel emissions. Some scientists and the health professionals believe it is very high and other health professionals think it is relatively low, and their answers are in a very wide band, so it is unclear what the correct number is.

CS: It certainly *is* unclear, but you must have been somewhat swayed by the recent Harvard Medical School report estimating the externalities of coal in the U.S. at $700 billion a year.

JT: No, I don't think it was a very good study. Anyway...

CS: But can you speak to that? I mean, granted there is a range...

JT: I don't think that is true. In other areas of science and health in particular, you don't find massive divergences in views.

CS: Well, precisely why don't you think that the Harvard Medical School report was a good one?

JT: The first problem is controlling for all the factors that are necessary to control for. Confounding variables are always a big problem in every epidemiological study, and it is not entirely clear that they did control them correctly.

Secondly, pricing these things is very difficult, and I don't think there is any ideal way to go about it. Take for example Cass Sunstein, who I think very highly of, but who doesn't agree with me on a lot of issues. He makes the point that those who estimate lost years when it comes to damage function often put a fairly low number for those who are in their 70s who die a few days early from particulate matter concentrations. So there will be a small number for the three or four days lost at the end of life; it's not that big a deal. But Sustein would counter and say people can traditionally spend a tremendous amount of money to stay alive for an extra month or two at the end of life when it comes to other things like heart disease or cancer, and willingness to pay ought to be our metric. And if willingness to pay is our metric, then the estimates we put down for damages for lost years in that scenario would be much higher than what a lot of economists are going to give you. That seems pretty reasonable to me.

So you can get damage functions all over the place based on whether you're going to look at willingness to pay or loss of productivity, and whose willingness to pay you are looking at. So the more you look at the underlying assumptions of studies like this, there are all kinds of problematic things. You can come up with very high numbers and you can come up with low numbers depending on how you do things in the weeds.

But I want to come back to the externalities issues because if we get into a tug of war about whether we should believe scientist A, health professional B, scientist C, health professional D, or a certain group of environmentalists and their favorite set of professionals to listen to, where those who are more skeptical have their own, we are going to get nowhere. Look at the universe of studies that has been done on these externalities estimates and just take the median numbers for the sake of convenience, which seem fairly reasonable to me, given the great degree of uncertainty. If you take these median estimates and work up from them, you find that the externalities costs are not significant enough, if you were to correct them, to really change things for them much. For instance, Bill Nordhaus, who is one of the most prominent economists who mucks around in climate change economics, calculates if you were to internalize the externalities associated with climate change when it comes to coal prices, it would increase coal prices by maybe a couple of cents a kilowatt-hour—not enough to make our renewables competitive against coal.

CS: I can't imagine how it is possible.

JT: Nordhaus is looking at median damage estimates reported by the IPCC. He is actually somewhat left of center. He is a consultant to the Clinton administration and he is not someone the so-called skeptic community is going to trot out as exhibit A in any sort of a public trial. But the point is that the damage estimates from climate change are relatively low in the near to mid-term and they will become very high in the future.

CS: Ahhh.

JT: So he takes the statistics from IPCC and says what will that imply as far as an internalization exercise? And that's what it implies. So the ideal internalization exercise would have a relatively low carbon tax present with an escalating figure as time goes on to account for that change in damage estimates.

CS: Well does that in and of itself make sense? You strike me as a man with a good heart, which means that you're concerned about what life on this planet is going to be like 50 or 100 years hence. Given that the damage we are doing now is going to have enormous repercussions in 100 years, shouldn't we be doing something aggressive about it now?

It strikes me that the person who proposed that solution doesn't understand that preventing a problem of global proportion is far easier than fixing it once it's wreaked its devastation. Either that or he simply doesn't care. His position is the equivalent of an oncologist encouraging someone with stage one skin cancer to continue to sun-bathe, on the basis that the condition isn't terminal at this point. If he were an oncologist with this absurd position, he wouldn't be practicing medicine very long, I can assure you.

JT: There are a couple of ways to address that. Thomas Schelling who is an economist who has played around with the climate debate, although it is not his primary area of expertise, wrestles with that argument and poses these questions. We are talking about significant damages from climate change that don't begin for about 100 years. If you look at the IPCC median estimates the damage seems relatively modest but after about 100 years you get some serious damage functions, if you accept median IPCC estimates.

So Thomas Schelling asked what if it were the year 1901 and we have all this information about what greenhouse gas emissions were going to do to the environment in the future and we had decided as a society in the year 1901 what the damage functions might look like and what the right remedies might be. How would we be thinking about this in the year 1901? His point is that the future is far less knowable than we imagine.

So imagine ourselves looking back in 1901. He said if economists were busily doing the best work they possibly could about damage functions from climate change, do you know what the number one damage would be? Mud. We would be primarily concerned about mud. Mud would have a huge impact on agriculture, transportation and most of the lost GDP would come from mud. And if we had some mitigation policy we would mostly be worried about alleviating the impact of mud.

He said now imagine also the year 1901 thinking about the technologies of the future. What would we promote? There is no nuclear power; gas-fired power was not particularly relevant at that time; we had a series of technologies that don't bear much on our present existence. His point being is that there would be no way to craft intelligently any sort of remedy in that period of time, not because people were less intelligent than we are now, but because society 100 years hence is unimaginably different than it was in 1901.

CS: The concept that climate change will have *de minimus* effect over the next 100 years doesn't seem to be shared by too many people. Having said that, I have to admit that your argument is strengthened by the fact that the rate of change for the next hundred years will be *far* greater than it was in the previous 100 years.

JT: Steven Landsberg then took a look at that same question. So the upshot from Schelling's analysis is trying to estimate what damages will be in 100 years from greenhouse gases is virtually impossible because we can't project the direction of technology; we don't know where we are going with energy; there is no way to do this intelligently.

CS: So we could develop fusion—hot or cold. We could develop anything.

JT: Who knows? It is impossible to know.

CS: OK, but is that, in your estimation, a good justification for inaction currently? We have 97% of climate scientists telling us that we're headed for catastrophe. Is the fact that the future is unknowable a justification for not doing anything here?

JT: It's not binary. I'm not saying that either we have tremendous revolutionary policy change that fundamentally reshapes our civilization or we do nothing; there are gradations in between. Schelling is pointing out that hitting the panic button and undertaking massive civilizational change to head off damages that will primarily come in 100 years and beyond is a fairly dubious proposition.

Then Steven Landsburg, an economist from the University of Rochester asked a different question which I think is equally relevant, i.e., whether it is morally defensible to impose major costs on the poor to provide better benefits to the rich. In any other context we would say no. But in climate change the fact that we are looking at generational issues, and that seems to be what we are looking at.

CS: How so?

JT: If you go back to 1901 in Schelling's exercise, median incomes, GDP per capita, human wellbeing, etc.— by any metric the average person in the United States in 1901 was in abject poverty relative to today. Would it have been morally defensible, looking back, for us to embrace a policy where these people impose major costs on themselves to alleviate the damages that will be visiting us in the year 2012 or something like that?

CS: Considering 1901, I think you have a point as it was mid-industrial revolution, and the economic gains from the common person were yet to come. I'm not sure that is as relevant today as it would have been.

JT: Well that's a good point; it depends on what you think that the improvements in human wellbeing for the next century might be like. If you look historically at trends since World War II, you see that we have a steady 1%—2% improvement in GDP every year, and if you spin those numbers out, in a 100 years we are going to be a far wealthier society than we are today.

One of my colleagues here at Cato argues that the improvements in human wellbeing that we have seen over the last 100 years are not necessarily going to continue because we have borrowed so much economic growth from the future for the present, in the form of deficit spending and all sorts of things that we may not see anything like that. We don't know what living standards are going to be in 100 years.

CS: To say the least.

JT: So that makes things very uncertain in calculations.

Landsburg makes the point that if we assume historical rates of improvement in human living conditions, then it seems pretty implicit that policies controlling climate change would impose by necessity major costs on the present to alleviate damages on the future which equates to taxes on the poor to provide benefits for the wealthy, which, in almost every other context, would not work.

CS: I am with you all the way until that last half of that sentence. So if we were to put together a massive effort to get this done along the lines of all the things that we have done in the past: the highway system, putting a man on the moon, building the internet—why would the cost of that be borne primarily by the poor?

JT: To reduce our reliance on fossil fuels and increase our reliance on renewable energy is to reduce our use of the cheapest source of electric energy in the market and to increase our use of the most expensive sources of electricity and energy on the market. That's going to necessitate an increase of energy prices. Energy consumption is a major part of economic growth in the society; it is a major aspect of individual expenditure; it is a major cost for people—and it is something that wealthier people can afford. If your electricity bill increases by 20%—40% for whatever reason, you will be able to afford that, but poor people will find that much more difficult.

The idea that we can get around that increase in energy cost by spending a lot of public dollars to affect the technological breakthroughs that will then make renewable energy less expensive and then cost won't be borne at all, and we'll all be happy is nice, but that's kind of a faith-based initiative. How much have we spent trying to make nuclear fusion economically better? $60—$70 billion in the past 50 years. Has that subsidy made any difference and preference made any difference? No, it is still the most expensive source in conventional electricity.

CS: Let me ask you this: it seems that the cost of all this stuff is plummeting, with PV way under $1 a watt, and the levelized cost of wind very reasonable. It doesn't look outrageously expensive. What am I doing wrong?

JT: The idea that we know what solar power will cost 5 or 10 or 15 years from now or what wind will cost, or what market share they will have, or what the technologies could be, with or without government intervention, is the purest of guesswork. There is a tremendous book I have on my shelf by Vaclav Smil called <u>Energy at the Crossroads</u>. Smil's a polyglot; he has PhDs in four or five different fields and he has a chapter in this book which was published by MIT in 2004 called "Against Forecasting," in which he looked at the history of energy forecasts from futurists from energy firms, to think-tanks, to academics, to banks, to the Department of Energy, to trade associations, to advocacy groups, to the CIA – to virtually all parties that he could track down forecasting documents. And then he looked back at these forecasts to see how accurate they were and which parties seemed to have the best track-records. And to sum up his conclusions, drunken monkeys could have done better than any one of these people. The energy industry is best characterized by the fact that future progress in market share is simply unknowable.

Nobody has done a very good job with this. You can track back Amory Lovins' projections in the 1970s about where renewable energy would be at various points in the future and they were widely off. But it is not just Amory Lovins; you can go back and look at ExxonMobil's calculations in the 1970s and *they* were widely off. We have to be a little bit humble about our ability to peer into our crystal ball and say this is clearly the technology of the future. How do we know this?

That's what people were saying not just on the right in the industry, but on the left and the labor movement about fission in the 1950s. In the late 1940s the New York Times had an editorial talking about how coal-to-liquids technology was on the horizon; it's a new thing and we are going to get rid of oil wells and we are going to provide all of our oil needs from coal. Nice thought. It is certainly doable; that's how Nazi Germany helped fight through World War II. But it is no more economic today than it was then.

People made grand predictions about ethanol, and it has been breathlessly promoted for a long time now—and it is *still* more expensive than gasoline. On the other hand look back on positive changes. In 2006 did anybody believe that natural gas prices were going to plummet? No. You had Dow and Dupont leaving the United States and relocating facilities abroad where natural gas prices where cheaper because they saw no end to the spiraling increases

in gas costs. Nobody saw the revolution of hydraulic fracking and what that would do; it caught everybody by surprise. It was probably one of the biggest changes in the
electric industry in the last 50 years—totally a bolt from
the blue.

CS: So your point from the beginning of this conversation to now is essentially that even Smil, one of the most
highly regarded people on the planet in terms of energy,
is saying we have this "epistemic arrogance," i.e., that we
think we know what we are talking about but we really
don't know. Or, better put, that our ability to forecast the
future is extremely bad. But he concludes that we need to
do something.

JT: Sure he does. Smil makes this observation, but he
is also concerned about climate change and he believes
that society should take out an insurance policy to hedge
against possible bad events in the future from greenhouse
gas emissions. And that is a legitimate argument. But it
is a *different* sort of argument than what I frequently hear
from the left, which is absolute certainty about what future technologies in the energy industry would look like.
They're full of absolute certainty about damage functions
and the consequence of climate change 50 or 100 years
hence. We don't know if it could be great or it could be
small. There is too much uncertainty involved right now.
They are certain that we should throw money at technologies or industries that we would like to see affect

technological breakthroughs. None of those certainties are warranted in the slightest, not at all.

CS: Well I think if we had a statistically valid sample of the climate scientists who weighed in on this, most of them would say you are right: this planet has never been through anything like this before, at least as long as human kind has been on it. So you are right; there is no way to know if New York City is going to be underwater in 80 years. And Smil's comment about the insurance policy is one that most people find compelling—the fact that uncertainty is actually a reason to move, just like not knowing if you're going to be in a car accident is a reason to buy car insurance.

JT: When you say most people find compelling do you mean most average Americans or most experts?

CS: I mean informed people – both experts, and people who read what experts write.

Here is the way I look at this thing as an observer. When I started watching climate change a few years ago, most college-educated Americans were aware that global warming was a central problem for human kind and a threat to civilization. Now, this is largely split along party lines; I read recently that most college-educated Republicans think it is a hoax, think Al Gore is a liar, and use every snowstorm in Washington as proof. That is what I observe, but I am here to learn, so please continue.

JT: Well a couple of things. Look at public opinion data from Rasmussen and Gallop which have routinely engaged in this arena, and have asked people how much they are willing to pay to reduce greenhouse gas emissions. By the way, that's a different question than whether you believe that climate change is an existential threat. When you get down to brass tacks, how much are people willing to pay?

People are willing to pay virtually nothing. Half or more of the respondents, depending on the year the poll is taken, say they'll pay nothing. They don't necessarily believe it's a hoax; they just don't want to pay anything to deal with it. Those who are willing to pay anything will pay no more than $20—$50 a month of additional costs. Different polling firms, in different years with different ways of asking questions to get the answers come up with different figures, but the common denominator here is that people are willing to pay very little.

This doesn't surprise me, putting aside whether they should be concerned or not concerned; the one common denominator about public opinion when it comes to government is that people are generally not interested in incurring the costs today to alleviate problems of tomorrow. And that applies to anything—whether we are talking about the unfunded liabilities of Medicare or the defense budget.

Look at 1936-37. Hitler is rising in Europe; Churchill and others argue we have to hedge against this threat with increased defense spending in order to prepare for a war we hope we don't have to enter, but people are saying no, we are not going to do that. So even when you had this immediate threat, people said no.

With mathematical certainty I can write out what current obligations we Americans have as far as the fiscal imbalances of Medicare, social security, and so forth, but people won't do anything.

CS: I can see that.

JT: So this idea that in this one area that we are going to find an exception to the rule, I find is very unlikely.

CS: Well, there are a few issues at stake here. One is Americans' extraordinary mistrust of government. That, by the way, isn't shared by the Europeans, who pay a great deal for government, but have come to trust it to act honestly on their behalf. The other is people's extraordinary discount rates, what costs us pain now is far more important.

JT: At least in the political sphere, versus their private lives; in people's private realm their calculus is a little bit different.

CS: That's true.

JT: So that is the first thing that I would say about your observation that most people believe they are very concerned. Most people say they are very concerned, but do they vote that way? No. Do they respond that way when polling asks them these theoretical questions about what they think? No. So I am not all that convinced that people are anywhere near as alarmed about climate change as simple questions like "Are you concerned about climate change?" would suggest.

As far as the climate scientists are concerned, just to clarify what we have argued here at Cato, our climate scientist here, Pat Michaels, who is a big name in the climate domain, does not argue that climate change is not happening, and he is not arguing that industrial emissions have nothing to do with it. In fact, he believes that industrial emissions are the primary, though not the entire reason, for the global warming of the environment and for the changing climate.

Pat's argument is that all of the evidence we have in front of us suggests that the climate is less sensitive to industrial greenhouse gas emissions than many of the models assume. And he comes to that conclusion by looking at observations with regards to temperature and various things. It is just that the models are not playing out; the climate is just not as sensitive as some people think it is. And if what we are seeing today is indicative of what we are going to see tomorrow—most models actually show linear degrees of warming, not exponential degrees, then it's going to be a relative non-event. Climate change will impose some costs on some; it will provide benefits for others, and the net impact is going to be relatively invisible to society as a whole.

He may be very well wrong, but I just wanted to establish that, because I think a lot of people who might be sitting here talking to me would assume that the people at the Cato Institution are arguing something in the climate change arena that we are not.

CS: Well that's gratifying.

In your heart of hearts do you believe that all of this discussion is essentially fair-minded? I am sure you are thick-skinned and I am sure you are asked this all the time, but this is an important question. I have about 9,000 subscribers to my blog, and a lot of them wrote in when I said I was going to speak to Jerry Taylor, they say ask him this: Is the Cato Institute's approach to all the stuff that we are talking about here really honest science? Or is there some agenda here? Originally Cato was co-founded by Charles

Koch, who obviously has an agenda. Is this really good, pure peer-reviewed science, or are we trying to prove what we have already assumed? Can you speak to that? Just don't punch me.

JT: No, no. The flip side of that coin is what I hear often from colleagues or friends who are conventionally thought of as right of center. Are environmentalists concerned about climate change for honest reasons, or are they quasi-Marxists who have long wanted to shut down the industrial society, and this is just the latest rationale of many? We were supposed to shut down the industrialized civilization because of disappearing oil deposits and that turned out to be a hoax.

CS: Peak oil?

JT: Yes. Peak oil is nothing new, if you remember the 1970s Jimmy Carter argued that by the year 2000 the lights would be going out across the globe from the disappearing crude oil.

We were supposed to shut down society because of the massive extinction crises were going through, and that looks dodgier by the day.

People on the right say the same thing, and that is why they dismiss environmentalists. Well they say these guys are quasi-Marxists; they have long hated industrial society;

they've spooked us with a half a dozen stories about how if we don't go back to some pastoral pre-industrial world we are all going to blow the planet up. And this is a whole bunch of crap; these people simply don't like capitalism. Why should we listen to them?

So questioning the other party's motivations and honesty and true intentions is a game that both sides play to tune out the other.

So if you are asking me whether *my* work is honest I would argue that I try to be. In the past I have argued against oil subsidies; I have argued against coal subsidies; I have argued against nuclear power subsidies. There are people who give money to the Cato Institute who are in these industries and we make the case we make because we believe it is correct. The Cato Institute argued against both the last Persian Gulf War and the first Persian Gulf War and it cost us a lot of support, especially when we argued against the first Persian Gulf War. The Cato Institute has been arguing about getting out of Afghanistan last Tuesday, and *that* cost us a lot of support.

If you look at the Cato Institute, you'll find that we call them as we see them and it doesn't have a lot to do with where the dollars are coming from.

We may be wrong, but the idea that we have a predictable set of policy opinions based on what the oil industry thinks or what the Chamber of Commerce thinks is demonstrably incorrect.

And secondly, even if it were correct that we were paid puppeteers, of, say, Koch Industries, so what? You can have bad motives and be absolutely correct and you could have good motives and be absolutely incorrect. And basically we are going to have a conversation about whether the Greeks were right to say that *ad hominen* was a logical fallacy. Well of course it is a logical fallacy. It doesn't matter what my motives are if I say $2+2=4$ and I am the meanest bastard in the world and if you say $2+2=5$ and you are the nicest guy in the world, it doesn't mean that you are correct.

It is a short-handed way of trying to not engage in the discussion and say well the guy who has the best motives is probably correct. Maybe he is and maybe he isn't.

   CS: I'm not sure. Unprincipled people cheat and lie, and cheaters and liars tend not to present themselves honestly, and therefore tend not to make good scientists.

Having said that, you *do* have a point. A scientist friend of mine likes to say that when politics becomes senior to science, regardless of the direction it's being pushed, then we are all in deep trouble. In other words, when honest people who study this stuff and have spent their whole lives honestly trying to figure this out are under pressure to conclude that global warming is going to kill us—or that global warming is a hoax—that is when you might as well jump off a bridge.

JT: And my colleague Pat Michaels would walk in here and underscore that and absolutely agree with you 100% and go on and on about the corrosive effects of politics on science. So I think that is absolutely correct and I agree with that 100%. But it is inevitable that when the government is heavily involved in a scientific arena or when a scientific issue becomes a political football, it will be politicized, whether you're talking about climate change or anything else.

CS: This is a very refreshing way of looking at it. So you would say that the Cato Institute is largely libertarian, in other words, get government out of there.

JT: Cato isn't largely libertarian; it *is* libertarian. Unlike some think-tanks, we have a mission to promote libertarian ideas in public policy, and to convince Americans that libertarian ideas are workable, they are morally and socially just, and they provide better answers for Americans' problems and promise better improvements in human wellbeing by all metrics than alternative ideas. That is our mission; that is why people give us money; we are a mission-orientated think-tank. But we try to hold the libertarian banner high and make our arguments in the most intellectually rigorous manner we can.

We try not to engage in hackery or anything like that, because we are involved in waging and winning the war of ideas. It's a long-term process, and I think in the long run, strong arguments beat out bad arguments. And if you have incentives to lead with weak arguments as they are compelling in the short-term, you are not going to win the war.

CS: I was a card-carrying Libertarian for decades and I have changed my viewpoints on these things, but I understand where you're coming from...

JT: By the way just as an aside: John Passacantando, the former head of Greenpeace, was more or less a Libertarian through much of his young career; in fact he worked with Allan Reynolds who was an economist. He became an environmentalist because he became alarmed about environmental issues. He didn't abandon his Libertarian roots; he just became an environmentalist. There are a surprising number of people in the environmental movement whom I have met who not only were Libertarians in the past, but still think of themselves in many areas of policy as Libertarians, and they make an exception on climate as the government needs to do something proactive.

CS: I was just about to ask you that: Do you believe that the private sector will do a good job in behaving properly vis-à-vis the environment? For example, do you believe that complete deregulation, the abolition of the

EPA and the Department of Energy, making this whole thing go away would be acceptable environmentally?

JT: I think it is unclear. There is one school of thought which would argue that it doesn't really matter whether business is thinking about "doing the right thing" for the environment or not. If we simply let industry do what they would do absent environmental regulations and controls, you would find improvements in the environment would follow—not because of any sort of high-minded collaboration with the Sierra Club or enlightenment on the part of CEOs or anything of the kind, but the very fundamentals of capitalism encourage resource economization, because that affects input costs and if you want to make a profit you want to reduce input costs. It encourages pollution reduction, as pollution is a waste, and stuff going up the smokestack can be used somewhere else.

An example of that is a Coke can. In 1950 the ability to crush a Coke can was a sign of virility. Today my grandmother can crush a Coke can; the amount of materials in that Coke can have been reduced dramatically. But it wasn't because of the EPA; it wasn't because of some collaboration with the Sierra Club; it was because they are capitalists, and they didn't want to spend so much money making Coke cans, so they found all kinds of ingenious ways to minimize mineral input, not only reducing the amount of resources that were needed to make a Coke can, but also reducing the amount of pollution in making Coke cans; there became far less scrap going through the system.

So there are many who would argue that simply allowing capitalism to go forward uninhibited would, at the end of the day, reduce environmental burdens because people have hidden and implicit incentives to do the things that environmentalists would like to see, i.e., minimize resources and pollution.

There are others who would argue that this has certainly been the case in the past, but it may not be the case in the future, depending on where you think technologies are going and to what extent we can expect improvements. They say that industry is not going to reduce greenhouse gas emissions unless they have a reason to – and they don't have a reason to right now unless they are hedging against the prospect of regulation in the future.

What you want to do is provide an economic incentive for them to do these things so that is where the whole internalization comes into play—whether through an explicit carbon tax or an implicit carbon tax, which is what cap and trade is. I think both perspectives are correct depending on what we are looking at.

CS: But that implies a government to impose those taxes.

JT: From a libertarian perspective we are talking about the environment as most people think of it; we are talking about a public good. Now there is a whole libertarian school of thought which would argue that we can privatize the public commons to a great extent and then treat pollution like a trespass.

In other words, we were talking about private property. Let's assume that I own the air over my house; it's not owned by the state regulatory agency, or the EPA or the local air resources board; each of us own that chunk of air over our property. Anybody who is polluting my air over my house is like throwing garbage on my back lawn; it will be like trespassing. We think of it in that manner when we are dealing with private property: pollution is a trespass, and it would be enjoined in a court of law; you have no right to trespass on somebody.

And maybe they will pay you for the trespass, like an easement. I may put a certain price on the air over my house and you make me an offer; maybe I will take the deal or maybe I won't; it depends how much they are going to pay me.

So from a libertarian perspective if you think of pollution as a trespass it is not something that you evaluate from the standpoint of the good of society. Libertarians are not utilitarian costs calculators; they care less about what is good for society as a whole; I'm not sure we even know how to measure what's good for society.

We believe in protecting the rights of people's life, liberty, and property, which means protecting them from trespass and pollution. When we are dealing with public commons where we can't privatize things – it is very unworkable for people to sue all the cars that are driving by and putting pollutants in their front yards. So we are dealing with public commons, what are we supposed to do? To argue that we should not regulate the public commons and allow people to pollute or to use it as they will is not a libertarian argument. After all, if we all own something jointly then we should all get to decide how that resource is used and how it is exploited and we might all decide not to have it polluted into the ground. But if you live in some parts of the rural south where the economy is terrible, and they need jobs and economic growth, they may decide to put up with an incremental increase in pollution because it is more important to have jobs here and some sort of income because they are living in poverty.

Each decision is defensible; none is morally better than the others; there are simply tradeoffs. So I don't believe that a libertarian world necessarily means that the public commons is left to fend for itself – not at all. So I think there is absolute defense for government to police the public commons. The issue is how best to do it, and at what level of government to do it.

The EPA has a definite role in policing the national environmental common, but does the EPA really have a good reason to regulate local commons, that is not a national commons? Say for instance air quality over Los Angeles. To the extent to which the air pollution stays in the LA basin, what is it to the people in Boston what tradeoffs are being made in Los Angeles?

CS: Well simply because the atmosphere is jointly owned, not literally but figuratively, by all seven billion of us.

JT: For some pollutants that is certainly true, if we are talking about the emissions which cross borders, absolutely. If I live in China, yes there are emissions in China that drift across the Pacific, so there are certain secondary impacts, so where emissions impact third, fourth and fifth parties, it is defensible to say that third, fourth and fifth parties should have a say about what goes on with the emitter. But where it doesn't, what business is it of theirs?

For example, take groundwater. If we are talking about the potential pollution or contamination of a groundwater aquifer, you can talk about the Ogallala, or you can pick any aquifer you want. The impact from that pollution is going to be fixed by the people who live above or use that aquifer, that might impact four or five states in the Midwest, or that might impact a small part of a county in Arizona. But if you live in Saskatoon, Canada it is nothing to you one way or the other; it doesn't impact you in the slightest – though you may have preferences for what you would like to have happen.

So there are a number of people like Alice Rivlin who now works occasionally with the Brookings Institute; she comes out of the Carter Administration, and she is a well-regarded left-of-center economist who argues that the best way to deal with pollution is to have the decisions made by those who are directly impacted. So if the direct impact for emissions is global, then it ought to be a global decision; if it's national then it ought to be a national decision; if it's state then it is a state decision, if it's local, then maybe the county may deal with it.

CS: There are going to be readers who, at this point, are going to say this guy clearly doesn't want environmentalism to move forward – that setting this up for an eternity of arguments about jurisdiction is just a way to make sure that it never happens.

To try to carve up which level of government—whether it is a city, county, state, nation, or international body—and then cross-referencing that against 80 different types of pollution…it's a legal nightmare. The reader is going to say this guy clearly doesn't care about the environment or he wouldn't be offering such an obviously unworkable idea.

JT: In other words the idea that state and local governments might in some circumstances be given the primary authority over pollution laws is evidence of lack of interest in seeing pollution controls imposed?

CS: Most of the readers are going to say that we have an international problem; it's a crisis that is affecting all seven billion of us, and expecting the good people of Hattiesburg, Mississippi to make the right call is silly in the extreme.

JT: Now when it comes to climate change it doesn't play out. We have to be careful when we talk about the environment and when we talk about industrial emissions or pollutions; we are not always talking about greenhouse gases; sometimes we are talking about particulate matter, or air toxics, or ozone. These are all different constituents and very different impacts.

So far the conversation is about climate change, so it needs to be a global decision. If we are talking about the toxic emissions from a dry cleaning facility it is probably local. If you said that talking about regulatory federalism in that regard is prima facie evidence that I don't care about environmental emissions and pollutions, that is nonsense.

If you look at what is going on legally today, what do you often see? You see state attorneys general suing the federal government for not regulating enough. Whether it comes to climate change or other issues you see this all the time. You see local governments and mayors filing suits against the EPA as well.

In other words, the idea that it is the federal government which provides the greatest promise in environmental action and it is the local and state governments that are the recalcitrant parties may have been true in the 1950-60s, even though I don't think it was; in any case, it is certainly not true now. The dynamic is entirely in the opposite direction.

There are all kinds of environmental statutes adopted in California which would never be adopted in Washington. You frequently now hear consumers saying the federal government needs to overrule regulatory initiatives undertaken by California and elsewhere in the interests of national commerce. I am very skeptical of those arguments, but some of my colleagues embrace them. In any case, I don't think that it follows that environmental regulatory federalism would necessarily lead to less aggressive regulation; I think it may very well lead to more.

CS: What do you think about the effort to establish a constitutional amendment to overturn the Supreme Court's "Citizens United" decision of January 2010, which allowed corporations the protection under the First Amendment to spend as much as they wish to influence our elections?

JT: I am very much against it.

CS: So you think it went down correctly?

JT: Absolutely.

CS: So let's talk about that for a second if we could. If you are ExxonMobil, and you made $41 billion last year, you can spend as much of that as you want to make sure that Craig is elected and Jerry is thrown out on his ear, or vice versa. In your estimation this is consistent with a concept of democracy as contemplated by the framers of the constitution?

JT: First of all I am not a specialist in campaign finance law. Put aside whether the decision was adjudicated correctly from a constitutional standpoint, because that is less interesting to me what the law ought to be as opposed to what it is. Corporations are simply groups of individuals who are engaged in commercial enterprise. They all bring with them rights of free speech.

A corporation is simply an aggregation of individuals engaged in commercial activities. Their rights to free speech don't disappear when they are engaged in commercial activities but reappear when they are engaging in some kind of noncommercial activity. There is no such thing as a corporation per se; there is an aggregation of people who incorporate for commercial purposes. If this aggregation of people want to jointly argue X,Y,Z, that's just like an aggregation of individuals that start a non-profit and argue for X,Y,Z. I don't see any reason to think that speech rights disappear when you are engaged in some activities but not in other activities, or if you earn your money in one way but not another way. Speech is speech, and the more speech, the better. It is very hard to make an argument that too much speech can destroy democracy or too much speech can be bad for the public polity. Speech is a remedy for bad speech.

CS: I see what you are saying. But it certainly seems that if we really believe in democracy as considered the way it was 237 years ago when the country was founded, we might think otherwise. This was a day in which, I'm sure you're aware, corporations were extremely carefully regulated; they needed to exist for an ad hoc purpose for a period of time and if there was no continuing purpose, they were dissolved.

JT: Just as an asterisk, by the way, there are some Libertarian legal scholars who believe that corporate law is a very dubious proposition which provides artificial protection for corporations that shouldn't exist; Richard Epstein is one of those. I think there is a legitimate argument about the means by which we regulate corporations and treat them in the tax code that is open to a reasonable discussion.

CS: So you are suggesting that their influence should be unlimited? I as a person have an extremely regulated influence in the democratic process. I can speak as loudly as I want, and I can buy my own television station, but there are limits beyond which I can't go, in terms of campaign contributions.

JT: Correct. There are limits that you can do as an individual with candidates, but there are virtually no limits which you can do privately. For instance if you are a casino owner in Las Vegas and you want to give $5 million to Newt Gingrich's Third Party PAC, you can give $10 million, if you want to give $100 million you can. So the limits placed on what individuals can do in the political arena are still there, but they are nowhere near as aggressive as they were 5 or 10 years ago.

CS: That is a good point. You strike me as a guy who honestly cares about the legitimacy of democracy.

JT: It's an interesting conversation we are having about renewable energy going in all kinds of interesting directions—which I don't mind, but my own opinion is that governments are responsible to protect the rights to life, liberty and property. That is what government's job is, and libertarians consider that is the most important consideration for public policy. Democracy is useful in that it allows us to elect people to affect that mission, protect life, liberty and property.

Democracy as a concept unmoored from that mission is not necessarily good or bad; it depends on what the majority wanted. Simply because an idea can get 51% of the vote does not necessarily mean that the 51% in question are going to be making wise decisions which are consistent with respect to defense of rights, liberty and property. I don't think it would be accurate to say that I am a staunch defender of democracy, because if you are a libertarian you believe there are certain areas which democracy should keep its hands off, like rights to life, liberty and property. 51% of the people *don't* have the right to tell you that you can't talk about x, or you have to use your property in a certain way, or that you have to live your life in a certain way.

In fact that is one of the big differences between the Cato Institute and the run-of-the-mill National Review conservative who constantly bitches and moans that the Supreme Court is intervening in areas that are properly left to the legislature. Wait a minute; the legislature is not supposed to be unbounded by any consideration for private right; 51% of the people just do not have the right to do whatever they want to do, and there is some places that are off limits to them. So that is my qualification here.

CS: What I am really talking about when I say democracy is this confluence of what the public wants and whether the public is allowed to do that particular thing, given that we have a constitution that is the supreme arbiter of how we govern ourselves.

JT: There was a case about 10 years ago in which Robert F. Kennedy Jr. and the NRDC went into North Carolina and initiated common law actions against hog farmers. Now they argued that the waste from these massive hog farming operations was destroying the ground- and surface water, and was causing a great deal of health harm, and also causing tremendous nuisance from the smell and from the other impacts of their operations. So they issued common law actions against these polluters. The hog farmers and their trade association said we have a right-to-farm law, which basically holds us immune from this stuff as long as we are complying with state law, which was adopted by majority in the legislatures and people can vote for these legislators – in or out. We have a right-to-farm law which allows a right to pollute up to a certain level, and whether it is harming groundwater or harming the surface water or causing a lot of smell or bothering neighbors or harming other third parties in other ways is utterly irrelevant. Democratic legislature gave us the right to do this.

And NRDC's response was no. There is a common law foundation to government which is designed to protect property holders from damage and trespass, and we are going to repair to that. Now if we are going to wax rhapsodic about the glories of democracy and how we must bow down to it in all circumstances, we tell NRDC to get the hell out of North Carolina. Democracy might mean a warrant to pollute, it might mean a warrant to trespass, it might mean a warrant to do great harm to individuals and the environment, because the majority is either ignorant of the harm or the majority embraces the harm and say it is necessary for economic growth and the industrial wellbeing. Libertarians would tend to side with Robert F. Kennedy Jr., and say that democracy cannot give you a warrant to destroy my property; government's job is to protect my property, not to facilitate destruction of my property. So I am greatly suspicious and generally hostile to repairs to democracy in conversations about environmental policy. What we should be concerned about is protecting people from trespasser harm and if there is a warrant for doing that, we should do it regardless of the democratic interest.

CS: I agree with you 100% here. The reason I got off on this tangent about democracy is that it seems to me the will of the people is being frustrated by a small minority of people who claim to have "rights" that I find extremely dubious.

And by the way, I don't think that science itself is a democratic issue. What's true is true, regardless of how few or how many happen to believe it to be the case.

But going back again to global climate change, the only reason I believe in global warming is not that I have any independent evidence; it's the same reason I believe that the earth is round or that I believe in the theory of evolution.

The only reason I believe in global climate change is because that is what 97% of these people spend their lives doing are telling me, and I have no reason to think that this level of consensus could be wrong on a subject that has received this level of inquiry from top-level minds.

JT: I think that is the wrong question. My colleague down the hall, the climatologist Pat Michaels, who has published peer-reviewed literature, is referenced by the IPCC, and is a member of the IPCC report. He believes in global warming as well; the issue is not whether or not you believe in global warming; the issue in that particular scientific arena is what impact climate change is going to have in the near future, in the mid-term, and the long-term. That is much more uncertain.

CS: Well yes, he is on the low end of that curve and somebody else is on the high end. So are we talking about a difference of degree and not kind? I don't dispute anything that you are saying but in the example of the hog farm, if there weren't a million hogs—if there were only a hundred thousand hogs, then we wouldn't be having the conversation in the first place. So there is a difference in degree and not kind? Is that what you're saying?

JT: I think that where we probably agree is that the right way to think about climate change is there are other reasons that people marshal for support for renewable energy but usually this becomes a climate change conversation relatively quickly. I would agree with you that the appropriate way of looking at the risk of climate change, having significant negative impacts on society, is to hedge against those risks—that you take out an insurance policy; it is reasonable; I don't disagree with that; that seems to follow the evidence, and I think people who are alarmed about climate change and people who are less alarmed about climate change can agree on that. Now the only real question is how big is the premium that I will have to pay for this insurance policy? Can I afford the kind of premium Al Gore wants me to pay or that CAP wants me to pay or should we be paying less of a premium for this policy?

And then the other question is of course is what will this insurance policy buy me? If you look at the Kyoto Protocol, it will reduce global temperatures by a trivial amount in 50 years and a trivial amount in 100 years. Even Tom Wigley, who is an alarmist scientist who is greatly worried about climate change, said that following the Kyoto Protocol wouldn't make much difference as far as temperatures are concerned. So that would be a policy that would have measurable costs which would have a negative economic impact, at least for some period of time, and it would produce benefits that would be virtually impossible to detect 50 or 100 years hence. It doesn't seem like that is a very good initiative. If you do the same exercise with the Markey bill in the last Democratic Congress before the Republicans took over in 2010 – the same calculations follows.

So you want an insurance policy where the benefits at least equal the premiums. And it's relatively unclear what that insurance policy might look like; I can't imagine what that might look like.

And I would then point out that right now, environmental concerns about climate change may very well be misplaced. If you look at the number which gives you the highest death count from climate change today, I think that 150,000 was the number that I saw in Science Magazine a couple of years ago. I think that was the largest body count figure I saw.

CS: Are you talking about extreme weather events?

JT: Yes. And of course not every extreme weather event has to do with climate change. There was an estimate in Science Magazine a few years ago that pegged the number of dead people around the world at around 150,000 a year, which pales in comparison to the number of people who die because of indoor air pollution, because they use biomass, and they don't have electricity, and they are living in poverty and that is environmental pollution which is killing many multiples of that 150,000. How many people are dying from poor water quality because waters are polluted because there is no water treatment? Many numbers beyond that.

You can go on and on and if you are going to put the dead bodies a year associated with just conventional pollutants it's already multiples beyond climate change, and it will continue to be so for the foreseeable future. So if we are going to take insurance policies, if we are going to worry about such things, it seems to me that we may want to worry more about the conventional pollutants and most of these conventional pollutants which are killing people around the world today are driven by poverty.

They are driven by the fact that they don't have electricity in a lot of parts of the world and they can't afford to spend the billions of dollars that *this* country has spent on water treatment facilities. We do know that the more advanced industrial base the less pollution you get from it. All of these things suggest that what we want is wealth creation, since wealth creation will have wonderful economic and environmental side effects.

CS: So let me ask you one final question on this subject. I am often asked what I would do if I were asked to market renewable energy, and I say that I would promote the concept of patriotism. It seems to me that most Americans actually care about the strength and viability of their country. They're coming to understand that our wars are about oil, and they don't like the concept that we're borrowing a billion dollars a day and sending it to people who hate our guts.

JT: Most of our oil purchases go to people who don't hate our guts.

CS: I don't find that argument compelling; every drop we buy from Canada means that somebody else is buying one more from the Saudis or Venezuelans.

JT: That's a fair point.

CS: So I just say that this is the most patriotic thing you can do, if you care about the health, and safety, and wellbeing, and expansion of prosperity of the United States, the very best thing we can do is to become a leader in "new energy."

I go back to what Patton said, or at least George C. Scott's rendition: "America loves a winner, and will not tolerate a loser." In less than a decade we'll be the second largest economic power of the world. We're not going to be happy there.

JT: People said the same thing about Japan in 1990.

CS: Even most of the right wing buys that.

JT: That is because they are trying to get Obama out of office.

CS: Good point, everybody has an axe to grind.

JT Everybody has an axe to grind on this national security and patriotism thing as well, and I must tell you I have absolute contempt for the argument intellectually. James Woolsey, Gal Luft and a group of generally neoconservative policy experts have long tried to reduce our reliance on foreign oil. Why? Do you think they went to an NRDC meeting and saw the light? No, they want to rain down bombs on the Middle East and they believe that our reliance on foreign oil inhibits American policy there. They believe that it constrains us; that we're worried if we were to do something about Iran getting a nuclear weapon that it would send the oil markets crazy and send the world off into an economic tailspin.

So their hope is if we can reduce the global reliance on crude oil, if we reduce the global economic impact from war in the Middle East, then we can engage in more of it. My boss here at the Cato Institute believes that if we were to reduce our reliance on foreign oil then we would reduce our interest in going to war there because a lot of people believed in 1990, for instance, that the reason we had to eject Saddam Hussein from Kuwait was to ensure that the global oil market didn't go crazy and he couldn't get a stranglehold on what consumers pay for gasoline. And/or that he wouldn't grow stupid-rich off all of these oil revenues and become some sort of global threat. Neither of those arguments were very compelling by the way, but they were arguments that proved compelling politically, and so his hope is if we were relying less on foreign crude oil we would be less interested in crusading in the Middle East.

CS: Actually James Woolsey is the second interview in my first book. He went into his point in which he agrees with Thomas L. Friedman's "Fill 'er up with dictators." In other words, if you look at the 40 really bad guys of the 200 sovereign nations, most of the worst are the ones that rely on a single commodity for the rent.

JT: What is the metric for bad guys? How many bodies are associated with these regimes?

CS: We are talking about totalitarian...

JT: I know; I am just curious as to what that means exactly.

CS: Well he didn't define it...

JT: Well of course he didn't. If we are going to define it by how many people per capita do these regimes kill, that is just nuts. The worst would be North Korea. What commodity do they have? What rent do they have?

None. So when you hear blousy statements like that they all sound very well until you kick the tires a bit.

For instance once I kicked the tires with Tom Friedman on this front, and he said that, obviously, the more money we send to the Middle East, the more money finds its way to the terrorists; and the more money that finds its way to terrorists, the more virile the terrorists are. So if we want to cut into terrorism, the most important thing we can do is reduce oil prices and thus oil revenues. It seems reasonable enough to me. I mean, all of that seems to follow, doesn't it?

But if you do a regression analysis and ask: is there really a correlation between oil prices (which are a good metric for oil profits) and terrorism? And terrorism based on either the number of terrorist attacks or the deaths from terrorists attacks. Well if Friedman were right there should be a correlation: higher oil prices equals more body bags or more terrorists attacks. Absolutely not. There is no relationship at all. And this is based on 20 years of data.

So I went to some dinner with the featured Tom Friedman, and he was at dinner with James Woolsey and the rest and he was beating that drum which he beats in every third column, and I asked him: how do we test this proposition? He looked at me blankly and said: what do you mean 'test it?' It's obvious; there is no need to test it. It's logic. I said, "Well a lot of times what you think is logical doesn't bear out. If we were to do a regression do you think we would find a correlation?" He said yes, and I said, "We have already happily done that for you and we can't find it. What do you make of that?" And then he just blustered and huffed and said it is irrelevant. So when you are in a world in which people don't consider evidence to be particularly useful, then we are in a world of religious belief.

I could go on and on about the patriotism of the security thing but I find that people who talk about patriotism here are intellectual scoundrels; there needs to be a better argument.

CS: Well I am all out of arguments, this has been wonderful. I knew this was going to be a winner and I thank you so much for your time.

JT: Well thank you I am glad you stopped by and I am flattered that you decided to include me in your next book.

(short break)

JT: You should talk to Jonathan Adler, who is a law professor at Case Western University. He makes a libertarian argument for doing something about climate change and his argument is as follows:

Even in that universe of so-called "skeptic" scientists like Dick Lindzen, Robert Balling, Pat Michaels and others they tend to agree that climate change is occurring and that industrial emissions are the driver; they just believe that the impacts will be modest and minimal. And many of them flirt with the argument that while there will be some cost from climate change, there will also be some benefits to climate change, and they will probably come out in the wash or come reasonably close to it.

But Adler says, even if we agree with that perspective of what we are saying here, i.e., that some people will be harmed and some people will be helped by climate change, so we should not be concerned about the harms because they will be offset by the helps. Well that is not what libertarianism is founded on. Libertarianism is founded on the role of government to protect your rights to life, liberty and property regardless of whether someone has benefited by violating your rights to life liberty and property.

The individual is paramount; he is not part of the utilitarian con. So Adler says: even if the scientists are correct, that is less interesting to me. If there are going to be some parties damaged by climate change then those who are doing the damages should pay the costs. And a policy which followed that trajectory sounds a lot to him like a carbon tax, it could even be a regulatory intervention, which would say we are not going to permit the harm because we can't do very much to redistribute your gains to those who lose, so we are going to prevent the harm in the first place.

CS: But isn't it very clear who is benefiting from this thing? If you have ExxonMobil, Chevron, Shell, BP, Conoco Philips, and so forth, it's obvious that they are hellbent on sucking the last drop of crude out of the ground and selling it to you and me, regardless of the consequences. There is no mystery as to who's benefiting here.

JT: Right. To me, that is the only real issue. I don't really have anything to object to with what Jonathan Adler argues. I think he is correct: the problem is having an effective policy which actually allows him to do such things. The difficulty in separating those who are benefiting from those who are harmed by climate change is pointed out by Cass Sunstein who works for the Obama administration and he is a former law professor from the University of Chicago—definitely someone who is thought of as left-of-center. He argued in a paper called "Climate Justice" with Richard Posner that most scientists believe that the bulk of the harms from climate change will be visited upon third world countries and equatorial regions.

Not all scientists agree with that, by the way, but that is the reigning belief. Well he said these people are also benefiting tremendously by global capitalism and wealth creation and international trade and technological advance. In other words, if you were to effect a world in which industrial emissions were kept at a sustainable non-climate changing level from 1901 to the present would wellbeing in India be better? Hell no, it would not.

So his point is that the benefits from industrial emissions and greenhouse gases also were visited on the victims. So any fair calculus would have to try to work all this out. And he said that this would be hopeless. There are universes of benefits and costs from climate change, and even in mundane situations they make it very difficult to execute the very simple and unobjectionable exercise that Adler suggested is warranted.

CS: There is no doubt that this is difficult—even if you were honestly trying to figure it out, which I personally doubt that most of these people are. Going back to what we were talking about an hour ago, I certainly believe that there are motivations—probably on both sides—that transcend the purity of just trying to figure this out and doing the right thing.

JT: There may very well be. There are certainly financial motivations that drive people, there are ideological motivations that drive people, there are careerist and social motivations that drive people, there are all sorts of motivations. I think it is largely a myth that there are people who are uncontaminated by external motivations which might impact their thinking. We cannot find that universe of people that we could put in a room to decide, and even if we could, a purity of motivation does not ensure a purity of thought; this goes back to the Greeks and ad hominem; motive is not a good metric for who is correct and who is incorrect. Nor is citing authority a good argument; the fact that most climate scientists think X is interesting, but it is not dispositive; most of them could be wrong. Scientists have had consensus on things and been wrong many, many times.

To me there is an intellectual laziness in these climate change conversations; there is a firm embrace of logical fallacies that are marshaled as if they were Euclidian certainties. There are accusations of motive which are blind to the fact that your own motivations can be equally questioned by the other side just as legitimately. You can question the financial motivations of the right, and the right can question the ideological motivations of the left, and you will get nowhere.

To a large extent, it is human nature for people to demonize those who disagree with them, and single them out and find convenient rationale for why they should pay no attention to them. I think well-thinking, well-meaning people should try to overcome all of those very understandable human instincts as best as they can.

CS: Thanks again.

JT: All right. Thanks, Craig.

Notes:

Insofar as this was an interview and not a debate, I didn't really "go after" Jerry too aggressively, even in matters in which I have huge disagreements. In some cases I may have made my dissenting opinion clear, but let me close here with a few quick points:

1) Jerry believes that our civilization is not duty-bound to take preventative measures against climate change because we don't have adequate visibility into the future. This does not hold water with me. Yes, we could be saved by a great number of things, e.g., a new technology or some unforeseeable event in the cosmos. But society's depending on the unknown to halt the destruction of our environment is not sane, responsible behavior.

2) Jerry argues that, since the greatest damage from climate change will happen many decades hence, our imperative to mitigate that damage *itself* comes decades hence. This is a similarly unsupportable position; it has no more validity than an oncologist who discovers a small tumor on my lung but does not advise me to stop smoking, since the greatest part of the damage has not yet materialized.

3) Jerry asserts that free-market capitalism represents a self-correcting mechanism that minimizes environmental damage because capitalism abhors waste. Again, this is specious. What capitalists abhor is wasting money, not

CO2, small but lethal quantities of heavy metals, etc. The choice here isn't between wasting harmful byproducts of fossil fuels or not wasting them; it's between cleaning up the waste or not cleaning it up. We have adequate proof over the past two centuries that, when industrialists are unregulated, they most often choose not to clean up after themselves.

4) The legal remedies that Jerry suggests are rooted in Libertarianism, a worldview that, in my opinion, offers fair and just solutions, and makes a great deal of sense in certain circumstances. Here, however, it's clearly inadequate. Litigating against polluters for "trespass," as Jerry suggests, will create a legal morass that the polluters (and their lawyers) will love, while the rest of the world slowly chokes and dies.

5) While I have to admit that Jerry's ideas about the *ad hominen* logical fallacy are interesting, it's tough to maintain that people of low moral quality make good scientists, or should be trusted to characterize the findings of science fairly and honestly. This doesn't even need to be confirmed with empirical evidence; it's true by definition: liars don't tell the truth. Unfortunately, this clarity of understanding eluded me in the moment of the interview.

Stephen Lacey—
Climate Progress

A t the time I conducted this interview, Stephen Lacey was a reporter/blogger for Climate Progress, where he wrote on clean energy policy, technologies, and finance. Before joining CP, he was an editor/producer with RenewableEnergyWorld.com, which is where I first met him many years ago; in fact Stephen also appears in my first book ("Renewable Energy – Facts and Fantasies"). I have to say how impressed I am with Stephen; a young man, probably half my age, he learned this industry backwards and forwards in an extremely short period of time.

Craig Shields: Good to see you again.

You mention that you personally have undergone something of a transition, and that you have gotten to the point where you are more cognizant of the extreme power of the political landscape in terms of where renewable energy is going. Could you expand on that please?

Stephen Lacey: Sure. I have come to realize that this is inherently a political transition. Even though the largest businesses in the world recognize that investing in clean energy is an important driver of business in the 21st Century, you are changing 100 years' worth of investment in infrastructure in a political system that has evolved around cheap fossils fuels. So you have this inherent political resistance to that transition—no matter how positive this story is on the business side.

CS: Now what I would love you to do, if you are willing, is to give my readers concrete examples. "Name names" to the degree to which you are comfortable doing that.

SL: I think one very common example that a lot of folks in the energy industry like to point out is Koch Industries. Charles and David Koch are among the richest men in America and own one of the largest private companies in the world which deals in chemicals, oil and gas production, and developing infrastructure on oil and gas. They've made it an explicit political goal to knock down any sort of state, regional or national promotional policies for clean energy or carbon reduction. They pledged tens of millions of dollars during the 2012 presidential campaign to defeat President Obama and within that have set aside millions of dollars for explicit advertising criticizing clean energy, saying that clean energy is too expensive, calling it "crony capitalism," etc. This is an explicit campaign; they are not hiding anything. They want it heard that they want to put millions of dollars towards bringing down this industry.

CS: When I see stuff like this, I have two reactions. First, I wonder if there is corruption underneath it; in other words, in exchange for those dollars, do they want something political to happen at the other end? But I also can see that this is simple business. Ford is trying to put GM out of business, and Coke is trying to put Pepsi out of business. One can expect that powerful forces will compete aggressively against one another.

SL: Yes, they have every legal right to put their money behind anti-clean energy campaigns, just like many of the environmental groups put their money behind anti-fossil fuel campaigns; that is American democracy; it's what we live with. At the same time, Koch Industries and other organizations like ExxonMobil and other conservative think tanks are part of an organization called the American Legislative Exchange Council. They charge members to sit down and write pieces of legislation which are then sent to state legislatures and often integrated without any changes. Many of them are anti-clean energy pieces of legislation, and if an organization is paying money to write legislation that is then adopted without edits, that is scary.

CS: Yes I agree.

SL: There are some very real political forces out there that are doing their best to stop the transition to renewables, and many of them see it as their political objective. They don't want government involved in energy, but many of these organizations rely on tax credits for their operations and have relied on 100 years of government support in the form of loan guarantees or indeed monies for drilling techniques, tax credits for manufacturing or drilling exploration—you name it. The government has been heavily involved in energy.

CS: E.g., the creation of the highway system.

SL: Absolutely. I think that you see a lot of these organizations latching on to the anti-government fervor right now and using that as a way to create a wedge.

CS: Yes, exactly. It is ironic, to me the unbelievable part of this is that these people have effectively taken the anger that has been rightfully generated by the common man against a force that wrecked the economy and then focused that anger against the federal government.

SL: I think we are continually trying to tell people in the energy sector that the government has been heavily involved every step of the way. Cheap energy is fundamental to the growth of this country, period. The government has recognized that ever since the dawn of the fossil fuel era. Now to somehow say that government shouldn't be involved in helping a new energy transition is disingenuous.

CS: Well let's take a step back. I am thrilled to be here and to see you in your new digs. I love the city for its architectural beauty and the potential that it stands for. Obviously I hate the dirty tricks and so forth, but tell me about the organization and its mission statement please.

SL: Sure. I work for the Center for American Progress, which is an organization set up to spread progressive values on the federal and state level. We work across a variety of areas: national security, economics, healthcare, LGBT issues, energy and immigration, etc.

I work on the energy team in communications, writing for the Climate Progress blog. We are an independent website based in the Center for American Progress, and we write news and opinions on the energy industry in general and on the clean energy industry specifically. We are writing about the science of climate change, the politics of climate change, the international project finance markets for renewable energy, the politics for clean energy—anything related to energy and spreading the progressive environmental views around energy. That is what we cover.

CS: That's terrific. And Joe Romm is the Editor in Chief?

SL: Joe Romm is the Editor. I read him for many years before moving over (from senior industry analyst at RenewableEnergyWorld.com)—and actually that is one of the reasons I did move over; he is very fair minded in that he wants to educate people about the science of climate change. He is very opinionated, he tends to have progressive values, but he is critical of people across the political spectrum who are either ignorant or are spreading misinformation about climate change.

So I decided when they had a job opening to come and join Climate Progress; he is great guy to work for and a very intelligent guy.

CS: I like the fair-mindness, and it's great to see intelligent people who are disdainful to people, regardless of their political tendencies, who don't do their homework and check their facts.

SL: Right.

CS: So what are the trends here? I'm reminded of a talk I had with Professor Ramanathan of Scripps, in which I asked him about public perception of climate change. I said that if we were talking about this a couple of years ago, virtually every college-educated American would have agreed that global climate change is real and it is an urgent problem, but that now this is split across party lines where college educated Republicans tend to think it's a hoax, that Al Gore is a liar, etc. He agreed that this has, in fact, become a political football.

Can you expand on what you think about this?

SL: Absolutely. A number of things have driven the recent dramatic shift, one of which is a variety of very intelligent conservative politicians trying to latch on to the anti-government fervor created by the Tea Party. When you see the shift in rhetoric from many of these leading Republicans who once supported legislation to combat climate change, who once supported the science of climate change and recognized the impact of greenhouse gas emissions, you see that their concerns are hinged on what government is going to do about the transition.

Many of them are fearful, or at least pretend to be fearful, about government overreach in helping develop many of the solutions, because ultimately when you look at the scope of the problem government has a pretty big role to play. Many people just don't like that; they don't want to face that fact. The extraordinary surge in the influence of the Tea Party has brought a lot of Republicans who once supported climate legislation away from that view.

CS: So you say government has an important role to play, tell me why you believe that? Why can't this happen in the private sector?

SL: I think it will happen in the private sector, but what you need to see is long-term consistency in the type of incentive that we provide for companies. When you have one- or two-year extensions of the only federal tax credit available for wind, geothermal, biomass, and hydro producers—and you have a *permanent* tax credit for exploratory drilling in the oil and gas sector, that is such a *gross* imbalance of the scales you need to remedy that in some way. Government has a role to play in remedying that.

Whether it is to make that tax credit on the renewable energy side longer term, or to strip tax credits away from *both* sources of energy, government is there to provide the rules for playing the game, and I think because there is this surge in interest in renewable energy and there is this extraordinary build-up of infrastructure in the fossil fuel industry, people are compensating for that. The government is investing a lot through loan guarantees, through the Treasury Grant program (known as "1603"), through research and development; there has been a surge in the last couple of years in renewable energy to try and make up for that imbalance, and that scares a lot of people; they are resistant to it because of the ramp-up in government interest in those sectors.

I don't know the answer there, but I am a firm believer that government needs to play a role in this transition— similar to what we have seen in other industries whether it be developing railroads, developing natural gas fracturing techniques, or public universities. We have set aside the incentives to make sure that businesses can flourish in those sectors that we see as strategic value. And renewable energy, clearly because of our environmental and national security issues, has to be one of our top priorities.

CS: Great. Thanks.

When I first started talking about "externalities" of fossil fuels, and quantifying the damage associated with lung disease or long-term environmental damage whether it is climate change or ocean acidification, people used to look at me with a blank stare. That's no longer true. To me, anyway, it seems that more and more people are starting to say that we need a way of creating some sort of level playing field that's built around paying the costs of these externalities. Do you agree, and where do you see this going?

SL: Absolutely, I agree. That's why we advocate for a price on carbon and other greenhouse gas emissions. Unfortunately in the current American, political land-scape we are just not going to get a price on greenhouse gas emissions. So you see the executive branch reaching out, and you see the Environmental Protection Agency doing what Congress couldn't to create a flexible pricing mechanism.

The EPA is now creating standards for greenhouse gas emissions for new power plants. They are using this very blunt regulatory instrument; rather than putting a flexible price on carbon where you can trade credits and provide the successful program that was shown to work under the Clean Air Act of 1990 for SOx and NOx emissions, the Environmental Protection Agency is just going to go in and fine you for a certain amount of carbon emissions that you are emitting. It is a very blunt instrument; it doesn't allow for much flexibility, and businesses don't like that. And the Environmental Protection Agency is trying to be as flexible as possible on the rules and they are trying to do it so it doesn't kill businesses. This administration wanted to see a flexible price on carbon; when congress couldn't do that, they are going to the Environmental Protection Agency which can form these rules outside of congressional action.

So the Environmental Protection Agency is now creating standards for greenhouse gas emissions from petroleum refineries, for new coal-fired and natural gas power plants, and these will, when rolled out, have a pretty substantial impact on new builds of fossil fuels infrastructure.

Now the reason why carbon prices are so important is what you alluded to, i.e., that the health and environmental consequences of our fossil fuel use are becoming increasingly clear. Tens of thousands of people in this country who live near coal plants are afflicted with asthma, costing our medical system hundreds of millions of dollars per year. You look at the economic impact of extreme weather associated with the build-up of greenhouse gases -- over $50 billion this year in the United States alone. We have the most extreme year for weather on record, and many climatologists and meteorologists say it is because of the build-up of greenhouse gases. We can now look at this and, though we can't pinpoint one and say that its cause was climate change, it is clear that the build-up of greenhouse gases created an extra punch in the weather, and that is fuelling some of the extreme droughts, flooding events, hurricanes, etc. in the United States and globally, and that it is causing hundreds of billions of dollars in damage. That is not valued in our cost of energy, as you know very well. So how do we value that?

It is a hard question. There are plenty of mechanisms for doing that: you can create a flexible trading system, you can create a blunt instrument through a regulatory regime like the EPA, you can create tariffs, try to price carbon on imported goods from other countries; there are a variety of way to do this. But what is clear is that we have to do this if we are going to transition to renewable energy -- particularly with the emergence of shale gas.

The natural gas industry has been calling gas a "bridge fuel" for the last 20 or 30 years because it is a cleaner-burning fuel and a transition away from coal into renewable electricity technologies. However, without a firm price on carbon, because of the dramatic dips in the price of natural gas we have seen, that is going to cause a flight away from investments in renewable energy because utilities are going to invest mostly in natural gas.

Now when you look at only the generation of electricity from natural gas, that is a good thing; you can transition coal plants to burn natural gas; you can develop new combined-cycle natural gas plants that will eventually phase out coal. That's a great thing because it's less intensive in terms of greenhouse gases. However, on the extraction side, there are still a lot of questions about how much methane and other greenhouse gas emissions that you are leaking. Methane is 25 times more potent than carbon dioxide, so at a leakage rate of between 2%—4%, when you are actually extracting the natural gas, studies have shown that natural gas on a lifecycle basis is dirtier than coal.

Now we still need to see the National Academies of Science studies on this; in fact, there have been a number of studies on both sides and the science is just not settled yet, so we do not know if natural gas is dirtier than coal. But given all of these questions, and given the dramatic price drops that we have seen in the natural gas sector, a price on carbon and other greenhouse gas emissions is *the way* to truly make natural gas a bridge fuel into renewable energy technologies.

CS: Let me ask you about a renewable portfolio standard on a national level. I know this requires some thinking, since each region of the United States has a varying amount of renewable resources; we have got sun in the deserts, you have got wind in the plains, geothermal in the mountains, etc., —and the southeastern part of the United States is not rich in renewable resources at all, and so maybe you need to make allowances for that. But I can't imagine what's the matter with a federal energy policy that says, "Ya know what? We're moving to renewable energy over a period of these next couple of decades." I know this isn't China; I know the American people would need to support the idea, but I think they would.

In any case, let me ask you what you would do if you had the power to lead our nation in this arena.

SL: I don't quite have the answer to that because I think there are a lot of smart minds out there with very different points of view on what is most effective for deploying renewables. Clearly we need policies that are going to help deploy renewables as quickly as possible and encourage efficiency as well, given the looming climate crisis.

But you can't just set policies for developing renewables and think that it is somehow going to transform society overnight. We have deep problems in the way we develop our infrastructure, build our buildings, drive our cars, build our highways; we can't just expect to develop a bunch of wind and solar and maybe some electric vehicles and then pretend everything is going to be OK. I see this as a fundamental restructuring of our society, and I think there are a lot of ways to go about that, but I haven't quite found what that answer is.

CS: When you say restructuring our society I am intrigued by this, and I agree with you but that opens so many doors and means so many different things.

SL: The decrease in suburbanization and the increase of urbanization as one. I think crafting more walkable community-oriented cities that will inherently reduce fossil fuel use is an imperative. I think this is probably an opinion that will get me in trouble with a lot of people, given the tens of millions of people that are living in the southwest, but I think to develop cities in the southwest where we have very few water resources, where we are going to see continued dust bowl conditions given the extreme weather events, where the amount of energy that it takes to truck food into cities in the middle of the desert, is just not a sustainable model. Over the long-term we need to start to think about how we are going to transition the tens of millions of people in the southwest toward more sustainable communities.

That means transitioning the way we have developed our communities in the southwest, on the theory that we are always going to have cheap abundant fossil fuels and cheap abundant water; that theory is completely incorrect, and that is going to hit us sooner rather than later, and communities are already making that transition.

So when you look at the way that we build our buildings, you see some major shifts in the focus on efficiency and on more sustainable materials. However the vast majority of buildings built in this country still use conventional techniques, inefficient boilers, somewhat inefficient windows, HVAC systems that provide temperatures that are different in every room, poor lighting, and we are still constructing our buildings in a way that doesn't necessarily mimic nature and doesn't provide the cost-efficient means of getting energy in and out; we still have a long way to go before we make that transition.

In transportation we are still dependent on petroleum and will be for a long time. We have seen some positive shifts in the electric vehicle market, but consumers are just not adopting them en masse. We still need to see battery costs come down. The unconventional cellulosic biofuels industry is making some steady steps but they are still in the early phases of commercial development, and we need a complete restructuring of our fuel system which includes new types of transportation, redesigning the car, making our cars smaller and lighter, completely redesigning engines, and redesigning our highways so we are not driving as much.

There are so many avenues we can explore, that I don't see how a simple focus on renewable energy is going to come in and solve all of our problems. That is one small piece of a massive piece, given that our society has grown up with cheap abundant fossil fuels and is now facing the limits of those fossil fuels.

CS: Brilliant.

The first interview in "Is Renewable Really Doable?" is Robert Pollin, an economist at UMass Amherst, and part of the ARPA-e committee. He says very much what you do and includes the idea that energy efficiency and conservation make use of extremely stable and proven technology. Part of the reason that people are not running to invest in renewables is that it is new technology; people are not investing in a lot of things and they are certainly scared to death of things that aren't proven. In other words, the real low hanging fruit is efficiency. Take passive solar, for example. If you're designing a house, can you make sure that you're getting as much sun in the winter as possible, and as little sun in the summer? Just take a few minutes and think about that.

SL: Right. We need to see efficiency as a resource. We are now starting to grasp the idea of thinking about efficiencies as a supply side resource; innovative demand response companies like Comverge and EnerNOC and Honeywell are thinking about how to pool energy efficiency for disparate sources into one resource they can then sell back to a utility.

They will go and they will contract with individual businesses in a particular utility territory or a particular city and say: when demand spikes during the highest points of demand in the year and the hottest days of the year, you will agree to shut off your refrigerators, turn off some lights in the back room, and do a couple of other things that add up across these disparate points on the system. And they get paid for that service; they get paid to power down those refrigerators, etc. The network operator pools all of them together and sells that back to the utility and says: rather than paying for that expensive new peaker natural gas power plant, we can efficiently utilize your system; we can pool all of those efficiencies and sell it back to you for cheaper than what it would have cost to develop a centralized power plant.

That is a concept that has been in the works for a while, but it is really catching on in the mainstream energy sector, and it's a fascinating think of using less energy as an actual resource. Once we get there, I think that efficiency will be appreciated for what it is worth.

CS: I hope you're right.

Here's a question for you that I have been asking everyone else. I know you read my stuff occasionally, and you see that I often write about the political stuff that underpins this transition. I frequently point out the folly of the U.S. Supreme Court decision of 2010, "Citizens United vs. Federal Election Commission," granting corporations the First Amendment rights of free speech guaranteed to real people. Can you comment on that?

SL: I don't have many terribly intellectual comments on Citizens United but I will say that because of the creation of super PACs donors can give as much money as they want to these political action committees and not disclose their donations. So you see many of these super PACs that support Republican candidates re tearing down clean energy and playing up the Solyndra debacle as an example of crony capitalism and why the government should not be investing in clean energy. Now in some cases the organizations putting forward these ads are up front and in some cases we don't know who is donating, and you are going to see this get more and more intense over time.

In terms of the energy sector I think it has enormous implications and I am scared about the amount of undisclosed money that will flow into the anti-clean energy campaigns; that makes me nervous.

CS: I can imagine. Speaking of money and campaigns, I meant to ask you earlier when I was asking about the mission statement, how is this place funded?

SL: A variety of wealthy donors and political organizations. I don't work on the funding side of things but we have hundreds of thousands of donators.

CS: Do you want to talk about where progressive action will come from in the energy sector? There must be some good news here somewhere.

SL: I think the story is the same today as it was five years ago and that is most of the action is going to come on a local and state level. This administration has continued supporting clean energy; Obama said in the State of the Union address that he will not back away from clean air despite the Solyndra loan guarantee debacle, despite how opponents have attacked investments in clean energy. The administration is going to at least rhetorically stay behind this industry.

Now there are a number of things happening outside of congressional oversight that will probably have an enormous influence on clean energy and efficiency. One is the adoption of vehicle efficiency standards by 2025, up to 54 miles per gallon, a dramatic increase in vehicle efficiency which will help to transform the automobile industry, increase manufacturing in this country, and help us reduce petroleum fuels and have an impact on our ability to look at next-generation transportation.

When you look at the phase-out of coal in this country and therefore creating room for more renewables and hopefully cleaner burning natural gas, the greenhouse gas emissions standards for new power plants that I mentioned earlier, the emission standards for natural gas fracking, water standards for natural gas fracking, emission standards for refineries, the mercury standard for existing and new power plants—all of these regulatory means are going to help usher in a transition to renewable energy. It spells a very difficult future for coal, and definitely opens up the market for more renewables.

So on the executive level there a number of things happening within the administration that could possibly open up the door for clean energy. The President has maintained support for R&D in his proposed budget; he wants to extend the expired Treasury Grant program for solar and wind and other renewables, he wants to extend the production tax credit, and he wants to increase R&D funding for the Department of Energy for next-generation technologies. This is pretty strong support. Now that budget will not go through, but it is a signal that there is some support from Washington.

However the actions really need to happen at the state and local level and we have over 30 states with renewable electricity standards, and some with renewable portfolio standards for heating, electricity and fuels; that will continue to play an important role.

Now some states, like Texas for example, are starting to reach the limits of their renewable energy standards. In Texas you see the development of wind energy drop considerably; they were developing a couple of gigawatts a couple of years ago and now they are down to a couple of 100 megawatts, and that is because they are almost reaching their targets. It's also very difficult to develop a merchant project in Texas because natural gas spot prices are so low so; if you don't have a long-term contract for electricity it is very difficult to just say you are going to develop a wind project and try to sell it on the spot market and compete with natural gas prices that are at $2.50 now. You may be competitive with $3.50 or $4.00 gas, but not $2.50.

You see programs like Property Assessed Clean Energy which allows a municipality to issue a bond, loan people money for energy efficiency of renewable energy retrofits, and those people pay back the city through reassessed property taxes or through different taxes on municipality waste. They find creative ways to factor it into your taxes. So that's a very interesting program that we have seen work quite successfully on the commercial level, up until the nation's leading mortgage lenders Fanny Mae and Freddy Mac stopped the program on the residential level because they had their hand in most residential mortgages.

They were concerned that the Property Assessed Clean Energy Loan would get paid back before the mortgage itself if the homeowner defaulted. There were a lot of concerns around that, and they decided that they were not going to support the program which pretty-much killed the program. There were a lot of efforts made to try and bring it back, to try to force Fannie and Freddy to participate, and we will see what happens. But that could be an extraordinarily important program on the local level.

One more thing: we have a lot of clean energy funds on the state level that have been set aside. So you are seeing about $250 million dollars a year spread across over a dozen states that have specific funds set aside for residential rebates for solar energy systems or energy efficiency retrofits, for business tax credits, and you are starting to see some of these states deploying these funds for broader economic development initiatives. For example, some states are creating cleantech clusters for R&D to try to capture the value, rather than just focusing on installation and providing rebates; they're shifting some of the money toward encouraging manufacturers to set up shop, to encourage business incubation, and to looking beyond the actual installation into materials, innovation and manufacturing.

So it *is* happening.

It's easy to get really upset about the gridlock in Washington and about what is not going to happen. But there is enough happening that even though we may see some significant problems in Washington, the industry will still march forward; I am confident of that.

CS: Why don't we close with that? How nice to see you again, my friend. Thank you.

Jeff Siegel—
Green Chip Stocks

J eff Siegel is the managing editor of Green Chip Stocks'
   "Power Portfolio," an independent investment research
service that focuses primarily on stocks in the modern
energy and infrastructure markets. Jeff also works as a
consultant, has been a featured guest on Fox, CNBC, and
Bloomberg Asia, and is the author of the best-selling energy book, *Investing in Renewable Energy: Making Money on
Green Chip Stocks.*

Craig Shields: Thanks for taking time for me here,
Jeff. I'm trying to "take a deeper dive" into the subject
of renewable energy, given that there is such a compelling set of imperatives to go in this direction, whether
your concerns are national security or health care or environmental issues – or even jobs and economic prosperity,
which I guess is what it boils down to for most people.

I'm trying to get a well-rounded perspective here; in fact, I was at the Cato Institute yesterday.

Jeff Siegel: I read that. I was curious to know how that went.

CS: It was very interesting but their story involves accepting some logical and legal maneuverings that I find kind of convoluted and ultimately doomed to failure. If you're a pure Libertarian (and I used to be one) you really have to work hard to think that the private sector will do the right thing by the environment and that individual property rights are all that are needed to keep pollution in check. They say, for example, that if you have a house and want the air above and around it to be clear, you say that a polluter is essentially trespassing. You go after a trespass.

JS: It sounds great in theory.

CS: Not to me. The idea that the private sector can be counted on to do the right thing by the environment is provably false. I'm glad I had that meeting because it really opened my eyes to ways of thinking about this.

In any case, if there are so many different imperatives why are we doing such a miserably poor job in this thing? I'm presuming here that the answer is essentially money. This new book is called *Renewable Energy: Following the Money*. What I'm trying to do is get as many different perspectives as possible on the incentives and disincentives for people to invest in this space.

I know that you follow a number of different specific equities but I presume you also follow the sensibilities of investors. In other words, what are they thinking? What are their fears? What are their aspirations and so forth? Do you want to just give me a little thumbnail on this since the last time we got together?

JS: It's interesting because I actually had this interview for a documentary a few months ago. This guy asked me a similar question and he's like 'Why has it taken so long for this alternative thing to happen? Why is it so hard for us to grow this thing?' I said 'Well, if you look at the numbers it's actually growing quite rapidly.' Look at solar, wind, geothermal, electric cars, even high speed rail if you don't count this country. The growth is there and it continues to be quite robust.

I don't know if the argument that 'it's taking a long time' is accurate. I imagine it could be a lot faster if, as you know and we've discussed in the past, our government weren't run by special interests. I don't know how many times we have to keep bringing it up that we've always picked winners in the energy game and losers in the energy game because we continue to subsidize oil, gas, nuclear and coal. Every time you bring it up it kind of gets brushed off.

CS: Right.

JS: Fox News or Rush Limbaugh and the rest of these guys talk about socialism and tax subsidies for electric cars. I never hear them say anything about the oil and gas industry. They won't even bring it up. They say 'Oh, well we need this stuff. These are incentives.' Last time I checked, when you have to pay the Navy to protect shipping lanes for oil, that's not an incentive; that's a crutch. I do think that the industry is chugging along quite nicely but it could move a lot quicker if these kinds of obstacles weren't getting in the way.

It's kind of like a hurdle race, when you have hurdles you can jump them or you can just move them out of the way. We could move them out of the way, but we have just chosen not to, so it's going to take us longer to win the race.

CS: And your suggestion is that special interests are at the root of this?

JS: Absolutely.

CS: Do you have any evidence to support that?

JS: A lot of times I look at the connections between the politicians and who funds their campaigns. By the way, I'm not anti-domestic oil production; I'm just making an observation. The observation is when people push really hard for domestic oil production, whether it be in Alaska or the Gulf, look at how much money they get from the oil and gas industry. Most of the time it's pretty substantial. These guys get millions of dollars in campaign contributions from the oil and gas industry. Do you think they're going to go against the oil and gas industry?

CS: This thing about the renewal of subsidies for oil did come up recently (March, 2012). The people who voted to maintain it received, per capita, five times as much in campaign contributions as the people who voted to overturn it.

JS: Right. I mean it doesn't take a rocket scientist to figure this stuff out.

CS: By the way, in the United States we were putting up a gigawatt or two of wind a year until recently. We went down to just a couple hundred megawatts last year. I am impressed with what is happening outside the United States. I don't hold out an awful lot of hope for what we're doing here, though that could change. Who knows?

JS: I think there are two reasons for that. One is there is still uncertainty whether the tax credits are going to be renewed. Obviously if I ran a turbine company I'm not ponying up if I don't know what's going to happen in the next six months. Another issue is transmission. There are a couple of transmission projects in the Midwest that are in development right now. The idea of putting up more wind turbines until you get that infrastructure doesn't really make sense.

That's one of the problems China has right now is they have wind turbines just sitting there in ghost towns. If you look at some of the big wind turbine manufacturers in China like Ming Yang and Goldwind, they're here in the U.S. Goldwind just did a big wind farm deal in Montana, and Ming Yang did one but I'm not sure where it was. That's a classic example of what I'm talking about: you put up these wind turbines but you should have infrastructure in place. It's such a gigantic waste of money, time, effort, and resources and plus you have a turbine sitting in the middle of nowhere for two years losing value. I understand when someone says, 'We don't want to build anymore wind farms until we have the infrastructure in place.' It makes complete sense. Those projects will be ready in about eight years.

There are two big transmission projects underway right now; they're not saying they are just for wind but this is where a lot of the wind is, so the assumption is that it is mostly for wind. It's in the Midwest and it's in Texas.

CS: That would tie that power into what population centers?

JS: The Midwest has gone into Chicago. Texas, I'm not really sure where the central grid is but I know they have been building transmission in Texas for a while. Texas is a huge wind state and another state that generates a lot of revenue from wind so they've been pretty bullish about getting that infrastructure in place.

CS: I understand uncertainty. The guy I spoke with yesterday, Stephen Lacey, an old friend from Renewable Energy World, who's now with Climate Progress, makes the point that uncertainty is not a friend of investment. Renewables have tax credits that can be turned on and turned off in a year or two; fossil fuels have subsidies that are written into the fabric of our law. Again, it doesn't take a rocket scientist to figure out where investors are going to feel more comfortable.

Can you give a more nuanced look into the psychology of investors here and abroad? We bemoan the fact that there are two trillion dollars of private capital sitting on the sidelines. What's up with that and where is that likely to go?

JS: As an investor myself, I wouldn't want to invest in something where I'm going to have an increased risk. From the perspective of some of these big money investors, we're talking tens of millions of dollars and a handshake. I wouldn't do it. I wouldn't invest in wind right now. I'd invest in infrastructure; I'd invest in transmission.

CS: What about efficiency? A lot of people say that the real low hanging fruit in getting us from where we are now to where we need to be, in terms of clean energy matching load, is essentially bringing the load down with conservation and efficiency. Are there things that you're following in that space?

JS: Not particularly. There was some demand management stuff that we were looking at a couple years ago but it kind of lost its luster. I think the whole efficiency thing has been underway for a couple of years now and it continues, especially in the building sector. You'll be hard pressed to find a new construction project now that's not using argon-filled windows. The world is following any kind of efficiency and conservation improvement it can make to that building. Some of it is because new codes require that and some is just because that is how you do it now. I don't know architects all over the world but the ones I do know, it's not even something that they have to pitch; it's just part of a deal.

There was a guy who was buying old properties and fixing them up and renting them, and he brought me in to consult on a couple of them. He had some clients who wanted to make their building green and energy-efficient. This one guy bought a beautiful three-story house in East Baltimore, overlooking the water. He wanted solar on the roof, so I said 'Okay, that sounds good.' I love solar just like the next guy but this guy's primary concern was going solar. We went into that house and they were still in the process of, based on what he wanted, making the upgrades to the furnace and stuff like that. He was picking the cheap water heaters and cheap furnaces.

I told him, 'If we put solar on the roof why don't you get yourself a super-efficient water heater and HVAC system?' I went online and showed him some of the air conditioners, the tankless water heaters and so on—I showed him how efficient they were and how much he could save on his energy bills. We went into insulations and special windows. When all was said and done and he spent all this money to put solar on his roof, but he didn't really accomplish anything because it was going out the window.

Fortunately, this is rare. Most people who are doing new construction or rehabbing these places, even just the windows, efficiency is common these days. I don't know anybody that doesn't put energy efficient windows in. I don't even think they sell windows that aren't energy efficient.

CS: Excellent.

Bring me up to date. The last time we talked you had certain stock picks that you were researching. I'm sure a lot of that has changed. What do you see as far as bullishness and bearishness in this sector? Are there any things that you like specifically that you don't mind naming?

JS: I think that the days of all these little solar companies coming up – when I first started if you saw the word solar in the name you'd buy it dirt cheap and it would move and you'd make money; those days are gone. Obviously solar has been shaken up to a good point. It needed to be. Competition is cutthroat now, and that's a good thing. When I look now at investing, at stock picks, I refer to it as the "modern" energy sector now because I don't really consider it "alternative" anymore. I'm looking at the big players now: Siemens, ABB, GE.

I mean look at a company like Siemens. Siemens sells everything; you can't walk into a building and not have one thing in there that wasn't manufactured by Siemens. Siemens is a huge offshore-wind turbine manufacturer. They are basically going to build most of the offshore-wind turbines in Germany. That's a huge deal. They do a lot of the HVAC systems; they pretty much do everything.

These are huge corporate behemoths; they can weather the storm. For wind, I'm more comfortable playing wind through a company like Siemens than I am maybe a pure wind turbine play. Our economic future is very uncertain and I'm not comfortable in these high-risk plays anymore. There are a few that I'll play that I think have a really good business model, or a really good technology. But most of my portfolio right now is large companies that are paying dividends, and if they go south this year or next I'm not too worried about it; I'm not going to lose too much.

These smaller companies, even mid-size companies, it used to be you could make ten or twenty percent over the course of a year and feel 'Oh, that's a pretty good gain.' But if we had a year like last year, boy, forget it: 40, 50, 60 percent down over the course of the year. I don't want to take that risk. I've never been a big risk taker and certainly when I talk to our members I'm upfront about it. I'm not a risk-taker. I will alert you to some trade you may want to make but me, personally, I don't. I did that years ago when it was a lot easier and safer to do it. Now, I just don't want to take that risk.

CS: I understand. Obviously we have the election year in the United States. We're just before the election, and this book won't hit the streets until 2013. But it's worth asking: What do you think, given the election year rhetoric and kind of voter sentiment and antipathy toward government, where do you see this playing out in terms of energy?

JS: I think that Obama is trying to get his "alternative energy mojo" back. He had an opportunity and he missed it. I think that hurt his credibility and I think he's trying to get some of that back now. What he does, if he gets re-elected, I have no idea. He does have an advantage now in that, with natural gas booming in this country, he can side with the oil and gas producers to support the natural gas development and I think that takes some of the pressure off him when he says 'Okay, I want to support wind and solar and electric vehicles.' That's my personal opinion. I don't know if that's true. That's just kind of the vibe I'm getting from it.

And of course, there is the natural gas boom; if he doesn't support it, he's going to lose votes. Republicans will slam him if he doesn't support natural gas, so he's going to support it.

If he comes out and says, 'I want to push solar and wind' and he doesn't say anything about the natural gas boom, he's an easy target—a gazelle in a jungle of lions. Like we were saying before, he's hedging his bets. I think he truly does want to push the modern energy, but he knows that he cannot do it if he doesn't offer an olive branch to the oil and gas industry. And the natural gas angle is pretty easy for him because there are a lot of benefits to having a robust supply of natural gas. It increases exports, it's used for trucks and buses, and so forth.

There are still a lot of environmental problems, especially with the shale. I don't know what the outcome is going to be with that but it's certainly not enough for him to take that political risk to avoid it.

CS: Let's talk a little bit about the European debt crisis. This morning the Greeks -- what actually looked very solid the day before now looks like a disaster.

JS: They should just cut them off. Give me a break. It blows my mind when I read all these articles. If I ran that show I'd say, 'Nice to know you. We're going to let that weight go.'

CS: My point is I'm amazed that places like Italy and Spain and Germany, not to paint them all with the same brush, but they seem to be moving forward pretty readily with a lot of renewable plays. Isn't that remarkable in the environment of this debt crisis?

JS: Yes, but realize that there, you have more support from the people. I think we're so manipulated here, more so than Europe. I've been to Europe a million times and I feel like I don't care if it's Fox or the Huffington Post – this is a perfect example. About a month or two ago Ford had recalled something like 600,000 SUV's and minivans. There was a problem with the braking systems or something like that and a potential fire hazard. Nobody talked about it.

CS: I see where you're going. The Chevy Volt.

JS: Exactly. I asked someone I worked with if they heard about the Chevy Volt that caught on fire? Of course they did, because it was everywhere. I said 'Let me ask you another question. Did you read about how the Traffic Safety Bureau — whoever did the final inspection and had to give their final results came out a few weeks ago and said that they determined that electric vehicles are no more or less safe than internal combustion vehicles? Did you read that?' No, he hadn't read that anywhere. Of course he didn't, because that doesn't sell. Electric car catching on fire. Hey. That's a great opportunity for Neil Cavuto to start talking trash again. Electric transportation is something he has no interest in being objective about.

I know when I talk to people, I don't care if they're liberals, democrats, republicans, or conservatives, most people have a skewed version of what's going on in the world when it comes to energy. When I go to Europe it's interesting. If I speak at a conference in the U.S., I have to start out by explaining some very basic things that are going on. When I go to Europe, and I learn pretty quickly, when I did that people look astonished that I was wasting their time pointing out the obvious.

There is a guy I work with, and the other day he was looking for a new car and he said, 'Yeah, I'd love to get a Prius. They get 30 miles per gallon. That's awesome but I don't want to spend 40 thousand dollars for a car.' I said, 'A Prius is $22,000. Why do you think it's $40,000?' He said, 'Really? I didn't know it was that cheap.'

It's not a big deal. Obviously I'm interested in this kind of thing so I'm going to be aware of it but why would you assume that it's so expensive? Because everyone in the media tells you it's expensive.

As you know, in Europe generally there tends to be more interest in sustainability. People feel, 'Well, of course you would do this to make the next generation have it a little bit better.' I remember in 1992 I was in Germany and I stayed with a friend of mine. We actually stayed at her folk's house. They had a recycling bin and a trash bin and I had accidentally thrown something in the trash bin that should have gone in the recycling bin and they asked, 'What are you doing?' They were not upset about it but they were shocked that I would do that.

Honestly I thought it was kind of cool that they had that because I had never seen that here at that time. That was in 1992 and that was a very common thing to do. If you look at the mass transit systems in Europe, the high speed rail, it's just a very common thing to embrace progress and not run from it. Here, we don't have that. I really find that a lot of people in this country, even if they say 'I want to do things that help the environment,' what they are really saying is 'I want to do things that are sustainable but it's too hard, it's too expensive.'

You and I both know that it's really not that hard or expensive to recycle or maybe to not drive everywhere. I live a five minute walk from the grocery store. My neighbor constantly drives to the grocery store. Really? Is it that much of an inconvenience to walk five minutes? Quite frankly, by the time he gets there and parks, he and I get there at the same time. These are just little things.

I really feel that there is this kind of misconception about energy and sustainability and it is perpetuated by the media. I think it's perpetuated by both the right and left. The right has an objective and the left is clueless. I don't know how you fix that.

CS: When you say the left is clueless, give me an example. There are all manner of organizations that are left of center — we mentioned Climate Progress earlier as one of thousands. I certainly wouldn't call them clueless.

JS: Yeah. I read their stuff all the time. I think they have some great stuff.

CS: OK, so please explain this, i.e., the left not getting this — are you saying they haven't been able to get their message out? I certainly agree with you if that's what you mean.

JS: Here's an example. From time to time I will speak at environmental conferences and inevitably someone says something along the lines of 'Why does it have to be about money? Shouldn't we be more concerned about doing what's right as opposed to making money?' My reaction is: 'What planet are you on? How can we do what's right if it can't compete in a free market?'

To me, that doesn't compute. I'm not saying that everyone on the left feels that way but I've heard this comment time and time again, specifically at these types of conferences. I get emails from people too. You and I both know that you can have a great idea; I could come up with a solar panel that could be super-efficient or it could be tiny and it could power a whole building but if it can't compete, if it's going to cost me more over the 20-year lifespan of that solar panel to have it, regardless of how wonderful it is instead of using natural gas, it's not going to work.

We get some of our power from coal, and I get it: coal *is* horrible. It's not competitive if we figure in the externalities. But we don't. Coal is more expensive than wind and solar if all these things were figured into the equation but they're *not* figured into the equation.

We have an uphill battle to climb. Instead of bitching about it we need to just climb the hill and get to the top. It's frustrating for me because I've had people sit there and call me a sell-out and stuff like that. My reaction is: 'What are you talking about? How is supporting a company that is working in the best interest of our environment, of our economy; how is that a bad thing?'

I often go back to the idea of what I do for a living, I do market analysis for modern energy companies, and I often make the economic argument in these cases. I already know the environmental argument and all these people at these conferences know the environmental argument. The environmental argument falls on deaf ears. If you don't know that by now then you're wasting your time. So make the economic argument because it is valid. Sometimes when you bring that up some people kind of revert back and say 'Why does it always have to be about money?'

I don't make the rules. That's how it works. If you don't think that's how it works I have a number of books that can prove to you that this is how it works.

CS: Having said that, these things do get more complicated when you get into subsidies. If we really, as a nation, want to compete in the international energy boom of the 21st century and if we really do care about the environmental consequences of what we're doing, we would put in place legislation that would level the playing field here. We should force our elected leaders to do the right thing or hit the trail.

But considering that the fossil fuel companies *do* have an agenda, and the left is clueless as you put it a minute ago—so considering that this is not happening what do you see?

JS: Let me rephrase this. How can I say this without generalizing? I know plenty of conservatives who agree with me on a lot of this stuff. Let's say you have over-zealous environmentalists that tend to be on the left. Then you have a percentage of those on the right who can't grasp the concept of progress. I have actually written about this before. I use this term a lot; 'partisan slaves.' They have no interest in knowing the truth. They have no interest in actually looking into an issue and making their decision. Whatever someone on their side says, they go with it.

This happens on both sides.

CS: OK, I'm with you, and I happen to agree, by the way. It's certainly true that the right has a status quo agenda, and that the left tends not to see the economic imperative for cheap energy. In other words, you can't sell something that is two and a half times the cost of electricity per kilowatt hour just because it's the "right thing to do."

JS: Right and you can't come out and say you're going to put a carbon tax on something because you want to limit carbon emissions. It's not going to happen. I don't care how many times you go to the White House and you picket. My personal opinion is 'Do you really think in these times, of all the taxes that all these people are going to submit, that that's the one that is going to get picked up?'

If you're going to take all that energy and time, why not put that into something that is going to get you a little further? Why not put that energy and time into doing a little research on the economic basis and make an economic argument so that when someone confronts you — and clearly people that are anti-alternative energy or anti-environmentalists — they love confronting you. They love the debate. So instead of countering that debate with 'Well, it's better for the environment' how about countering the debate with 'Well, it's better for the tax-payer.' How about countering that debate with 'Hey, this is better for job creation. We can create jobs.'

CS: Well, I was just about to go there. Would you rather pay someone to be unemployed or would you rather pay them to put up a wind turbine?

JS: Right. Use that money to create jobs. I mean you could do that with any number of things. It wouldn't have to be just wind.

But the job thing is a tough argument to support. Not that I don't believe it but anyone can come back and say, 'Oh, well if the government weren't supporting it we wouldn't have those jobs' and I could come back and say 'Well if the government weren't supporting the oil and gas industry then you wouldn't have those jobs either.' We'll go back and forth and no one is going to get anywhere.

CS: That's true. I've been there.

JS: I'm sure you have. But 30 minutes from here there was a GM plant that was closed down a few years ago. Now they are making batteries there. Did the government have something to do with that? Absolutely. I also know that there is a lot of domestic oil production going on right now. Did the government have something to do with that? Damn right it does. Let's not mince words here. Let's be honest about this.

CS: Yes. Exactly. Jeff, thanks very much for your viewpoints here.

JS: Happy to help.

Rajendra K. Pachauri, Ph.D. — Intergovernmental Panel on Climate Change

Rajendra Kumar Pachauri has been serving as the chairperson of the Intergovernmental Panel on Climate Change (IPCC) since 2002, which was awarded the Nobel Peace Prize in 2007. He has also been the director general of TERI, a research and policy organization in India, and chancellor of TERI University.

I met Dr. Pachauri at his office at Yale University, to which he had recently returned from a convening of the Conference of Parties, whose purpose is to arrive at a workable consensus about global climate change.

Craig Shields: What an honor it is for me to meet you; thanks so much for the time you're taking with me. I'd like to begin by saying that my first book, of which I've sent you a copy, contains an interview with a guy you must know, Professor Ramanathan.

Rajendra K. Pachauri: Oh, yes. "Ram" is a very close friend.

CS: I guessed that must have been the case. I spent a few hours with him a couple of years ago.

RKP: I can take a little bit of credit for having gotten him worked up about his project. We tried to do things together but he's been a hardcore atmospheric scientist. Yet he is a scientist with a very broad vision. He has been spending time with us in New Delhi and that's how he has gotten charged up with what we are doing; he is now very focused on solar energy.

CS: Yes. This was just a couple of weeks before Copenhagen. I remember he didn't want to express how fruitless he was afraid that was going to be but it was very clear from what he was saying. I could hear that he was working with other people to develop a backup plan in case it was the disaster that everybody presumed it would be.

In any case, I am working on my third book whose working title is *Renewable Energy: Following the Money*. I'm just trying to figure out if this is all so important why is it not happening.

RKP: I see.

CS: First of all, if you could help me sort out one thing. You've probably heard this metaphorical Plan A, B, and C where Plan A is we do nothing which is what we're demonstratively doing in the United States and just let the chips fall where they may. Plan B is we grow our way out of this; Amory Lovins, Jeremy Rifkin and the others write books all the time about job creation; you have mentioned "negative cost." In other words, the implication that we can go forward without a blow to our collective lifestyle. Plan C seems to be Bill McKibben and Nate Hagens saying, "The party is over. We're at the end of cheap credit and cheap energy."

Everybody can't be right here. Certainly Plan B and C can't coexist in the same universe. Where do you see this going internationally?

RKP: We have looked at renewable energy in depth in our Special Report on Renewable Energy Sources and Climate Change Mitigation at the IPCC. It's obvious that if you really want to influence the course of developments in the future we have to put in place a set of policies which will determine how much renewable energies will contribute to our energy production and supply. You remember that we examined 164 scenarios in which we got a range as low as 11 percent of renewable energy in 2050 versus 77 percent at the other end. I personally believe that you have to create awareness among the people.

CS: Among the people internationally? In other words, I would think this would be one problem in the United States and quite a different problem, in say, China.

RKP: Well, absolutely. You know, renewable energy is also not available uniformly. There are some countries where you have an abundance of resources, and others where that is not the case. I really think that we need to create awareness across the globe on the scientific realities of what is possible and at what cost. That, in my mind, will lead to several initiatives at the grass roots level. That in turn will influence policy. I really don't think one can legislate what has to be done at the national level or at the international level. I think the time has come where people should become much more understanding about the choices they have.

CS: And, again, internationally you're talking? Or are you talking about Americans?

RKP: No, I would say internationally, across the globe. Of course there are variations. There are some countries that are far more active in taking action where others are not.

CS: Europeans generally are active here.

RKP: Yes. Although, of course, right now they have some serious problems. There has also been a slight slowdown over there but on the other hand I think it is deeply ingrained in most countries in Europe that they will bring about a transition in the energy sector.

CS: Yes, exactly.

RKP: But it's also true that Europe doesn't really have an abundance of renewable energy resources that other parts of the world have. So therefore I think it is a case of horses for courses . We have to create that level of groundswell understanding by which we can get policies in the right direction. Unless you've got a mix of policies that will really bring about ways to use renewable energy sources, it's not going to happen by itself.

CS: Right. One thing that I talk about commonly is internalizing the externalities. In other words, if we pretend that there are no consequences for burning coal, we're actively encouraging our own destruction.

RKP: Absolutely. Well, internalizing these externalities, of course is an extreme outcome which will be very desirable, but on the other hand we have some very perverse subsidies on fossil fuels. So I think the first thing to do is to get rid of them and then you go to the next step in terms of internalizing the social impacts, the impacts on the global commons and all the externalities.

CS: Great. I did an interview Friday with a guy who works for the German Ministry of Environment and

Economics. I told him at the end, "I understand that you're doing this for Germany and there is nothing the matter with that, but recognize that if we can't collectively help China find a way to stop building a new coal-fired power plant at the rate of one a week, we're all going to lose." I think we certainly need to be a little more circumspect – perhaps less parochial in our way of thinking. Do you agree?

RKP: Well, you know, I think the Chinese know what they are doing. I've been interacting with them at the National Development and Reforms Commission; I see some major changes taking place in China.

CS: Well, they're definitely investing in a big way.

RKP: Not just that, they are also strategizing on how they can move towards a low carbon pattern of development. Now how much of that really takes shape and takes root is yet to be seen but the realization is certainly there at the highest levels of their government.

CS: That's fantastic. You sound more optimistic than I thought you would.

RKP: If I weren't optimistic I wouldn't be doing what I'm doing. Frankly, there are better things to do otherwise – No, I remain optimistic not because one wants to have a sunny view of everything that's going to happen in the future. But if you look at all of the developments that are taking place and put them together I think the sum total gives you some basis for cautious optimism.

CS: That's great. I watched your recent interview with Amy Goodman of "Democracy Now!" The thought struck me, since that show is viewed as "way left of center," it seems a shame that the traction you're getting is with the radical left. In other words, the fact that this has become bifurcated across political lines —

RKP: Unfortunately.

CS: What do you make of that?

RKP: Well, I expect that leaders of business will certainly see the benefit of moving in this direction and particularly in a globalized world when they get a clear perception of where the world is going. They're not going to stay behind. I'll give you an example. I just went to visit Sikorsky Helicopters today; I was very impressed with what they are trying to do. They are focusing very seriously on producing a product that is very low in energy consumption, that is inherently more sustainable. I was quite impressed and I found it very refreshing to see a company like that being so focused on these issues. In fact, I have a whole lot of literature that they gave me. You would think that this is some NGO.

CS: Yes, exactly.

RKP: But this (Raj hands Craig a printed brochure) is the kind of presentation that the vice president has been making to their top management.

CS: That's great. We're looking, by the way, at a photograph of two polar bears on a shrinking piece of ice in the Arctic.

RKP: What I'm saying is that it's just a matter of time that business will realize the benefits of moving in this direction.

CS: It seems to me that, ironically, business leaders seem to get this better than the politicians.

RKP: I think so. I actually made a statement that if business moves in this direction then clearly politicians will follow. It's just a matter of time.

CS: Yes. Well, tell me a little bit about your take on cleantech and job creation. You may know Robert Pollin at the University of Massachusetts at Amherst; he's a very senior economist, and part of his role is working on the APRA-E, figuring out how to spend this money. I asked him about job creation and where he saw the low-hanging fruit and so forth. Do you mind responding?

RKP: Well, frankly, I just don't have enough detail to be able to comment on this but there is a report that has just come out, I think a couple of weeks ago or something, where they have come up with estimates of green jobs which I have read rather quickly. I found it quite interesting. To be quite honest I would like to see a lot of analysis and hard data to be able to argue a position one way or the other. The indication seems to be that there might be more jobs from some of these new sectors of development but I'm not too sure if one can come up with a definitive statement on that. I think the jury is still out. We need to get much more evidence before we can take a position on this.

CS: Okay. Do you believe that there are forces of true evil – though perhaps that's too harsh a word. We read about the Koch brothers, possibly the most visible forces of maintaining the status quo approach to energy generation and consumption, even at the expense of the wellbeing of mankind. To a lot of people they represent a set of hidden forces that act behind the scenes to actively deter progress in this area. Do you agree that such a thing could exist or does exist?

RKP: Frankly, I don't have any evidence. In the Fourth Assessment Report of the IPCC, we have talked about a transition towards mitigation options in energy supply. That says that some of the barriers that vested interests will put forward will make this transition very difficult. The scientific community clearly understands this and I wouldn't be able to identify any specific personalities doing this work. This is inevitable. Whenever there is change there will be some who would question the validity of such a change and then others would do it for reasons of vested interest and we've used these words very clearly in the Fourth Assessment Report.

CS: I read things constantly that suggest that simply because we haven't had a disaster in the 100 thousand years or so that we've had a human species on this planet, that we aren't facing one now. All the people who predicted catastrophe were wrong; therefore the people who are currently predicting catastrophe are wrong. Well, that logic is specious. Having said that, you must run into this a lot yourself often, where people say, 'We've always worked our way through these things in the past without the Raj Pachauri's of the world.' How true is it that the sky is falling now?

RKP: Mahatma Gandhi came up with a very good statement. He said "There are two types of choices in a technological society. You can either indulge in self-deception and delusions until you are overtaken by catastrophe. On the other hand a culture could have certain inbuilt checks and balances by which you get over this self-deception and accept the truth for what it is before catastrophe overtakes you."

I would like to believe that in the end, human society will fall more into the second category rather than the first. In this day and age where we have so much intelligence, so much knowledge, so much instantaneous communication around the world, why do we have to wait until disaster strikes us? We have very clear projections of what will happen with the world's climate if we don't take action now. We brought all this together in a special report on extreme events and disasters; it came out in November 2011, and it's on the IPCC website. It is very clear that heat waves, extreme precipitation events, and other forms of disasters are increasing in frequency and intensity.

Now if we have that knowledge, should we not try to avoid that kind of outcome? Why do we have to wait until we become victims of it? That's what a risk-minimizing society, that's what a rational civilization would want to do. Therefore I don't see why we have to be victims of disaster or any kind of catastrophe before we are able to take action.

CS: Yes... the metaphor of fire insurance not because we think we're going to have a fire, we actually think we're probably not, but the consequences of the disaster are so extreme. Here, we know for a fact that climate change is happening and we're still not mitigating it.

RKP: My good friend, Steve Schneider, who died unfortunately over a year ago; whenever he was addressing an audience he would ask, "How many of you have had a fire in your home?" And if you have 200 people, maybe two people will put up their hands. "How many of you take out fire insurance?" Everybody put up their hands. So it's the same thing.

CS: Yes it is.

RKP: I think that's a very good example. As rational human beings we invest whatever is required in reducing the risk from some of these possible outcomes. That's what we should be doing.

CS: Speaking to this as specifically as you can, are there any late-breaking technologies that you think are important? Obviously different parts of the world have different renewable resources. You may have OTEC in the tropics and you may have CSP in the deserts and so forth. Are there any things that you think are, perhaps undervalued? Under-examined?

RKP: There is a whole range of things. We've said very clearly in our fourth assessment report that to launch a program of mitigation, *all* the technologies that we need to get started with are here today, or due to be commercialized very soon. Lack of technology is certainly not an excuse that holds any water. We can get started. Yes, for the future we need a package of policies by which new technologies and innovations will help us continue on the path of stringent mitigation. But we can embark on stringent mitigation today based on whatever technologies that are available.

CS: Great. In closing, let me explain that my website, 2GreenEnergy.com has about 9,000 subscribers, and I'm constantly asking these folks to submit business plans. Of the 1,200 or so who have done this, there are at least a couple of dozen that strike me as terrific concepts. I try to represent them to a community of investors, who themselves are a part of these 9,000 subscribers, i.e., bring good ideas together with people who can take them forward.

RKP: I see.

CS: I'm wondering, in closing on this thing, how I can serve you. I really believe in what you're doing and I will go to extreme efforts to support it.

RKP: That's very kind of you. I think there is nothing more important than spreading the message. I think we have to spread the truth; we have to inform the public about where we are with respect to the impacts of climate change. Where can we do things without necessarily giving up all the good things in life that people have become accustomed to?

CS: All right. I'll do that, to be sure. Thanks so much for your time.

Short Essays on the Subject

How Much Renewable Energy Does the World Really Need?

We occasionally see articles attempting to show that the goal of 100% renewables is closer than we think. At the risk of stating the obvious:

- We're a very long way from replacing coal using market-driven forces in places like West Virginia and the rest of the "coal-country" in America. The world is not running out of coal (unfortunately). In certain parts of the U.S., you don't even have to dig for it; you have to be careful not to trip over it when you walk through places like Wyoming. If we can't stop burning coal for other (i.e., moral) reasons, it's going to be around for at least a few decades, as the cost of wind and solar coupled with energy storage solutions slowly falls and the two graphs finally cross. Other parts of the world (like Beijing) have their own tragic stories to tell about the effluent of their coal-fired power plants.

• There are many places in the world where renewable energy is already the deal of the century, due, again, to purely economic factors. Did you know that Brazilian businesses pay $2.00 per kilowatt-hour for on-peak power? If you can't find a way to replace that with renewables, I don't know what to tell you. Putting this into perspective with the paragraph above, what they're paying is roughly *50 times* the cost of producing a kWh of electricity from coal in the U.S.

• Also, let's think about the island nations that ship in diesel to generate electricity at outrageous expense – both financially and ecologically. Let me clarify my use of the term "ecologically." Obviously, there is enormous environmental damage associated with shipping the dirtiest kind of diesel ("bunker fuel" as it's called) thousands of miles, burning it in 50-year-old power plants, and then dumping the resultant toxins directly into the atmosphere — but that's not my point. Rather, my point is that there are economic aspects to this practice as well, as prospective tourists to these enchanting islands and their dollars that would serve to drive these local economies no longer want to travel to places that have plumes of brown/gray smoke billowing into the air, coating the objects and people below with a thin but obvious film of diesel particulate. People are funny like that nowadays.

Another point in closing: the world's energy supply doesn't need to be 100% renewable. We just need to establish and maintain a steady course in that direction. This is something that is eminently doable. We simply need to care.

Notes on an Ideal Civilization

Last night, a friend of mine sent me a considerable laundry list of things he thinks everyone, certainly all U.S. citizens, are entitled to as basic human rights. I don't know. I see providing all this stuff gratis to a society that includes millions of lazy screw-ups as really distasteful, not to mention unworkable. Having said that, here are a few broad strokes that should be made in the direction of an ideal civilization as I see it:

Universal healthcare. Taking this out of corporate control will result in far better outcomes for everyone (except those seeking to profit at the expense of others' misery) — even for the doctors who are quitting in droves as more of them every day are realizing that what they're doing does not align with their personal sensibilities. Such a move will also immediately focus healthcare on wellness, versus the invention and encouragement of disease so it can be treated profitably.

Taxing wealth. The dominance of lawyers and other bullies over our civilization's last 500 years has resulted in the unfair concentration of wealth in the hands of a very few. Take Warren Buffett's advice. He's a smart guy, and he's absolutely right in this case in particular: If your legislators can't seem to figure out how to do this, fire them, and get people who can. It's not really that difficult. They work for you; you don't work for them.

Taxing polluters. I suppose I'm really something of aLlibertarian at heart, in that I really don't object to fossil fuel companies (or cigarette smokers, or helmetless motorcyclists, or prostitutes, etc.) as long as they pay the costs of what they're doing. Want to burn coal to generate electricity? No problem, just pick up the tab. It's $700 billion/year in the U.S. alone, and it's rising exponentially. If that cost isn't acceptable, simply stop behaving in ways you can't afford. The moment this happens, renewable energy will be heralded as the deal of the century, tens of millions of people will be hard at work developing and installing cleantech, and the climate crisis will be averted. And the figure I quote here doesn't include the cost of wars. Which leads me to:

Separating business from government. Get rid of "Citizens United" and the related legislation surrounding corporate lobbying that has all but ruined what our forefathers sacrificed so greatly to provide us. What will happen when Big Money no longer owns Congress? Actually, all the above-mentioned garbage will stop more-or-less instantaneously. No more wars to provide Big Oil access to foreign crude, no more protection of Big Pharma in its quest to addict every single one of us to some form (preferably several) of medication, no more favors for Big Food and its reckless indifference to our health (especially childhood obesity, diabetes, etc.) It will all be gone with the stroke of a pen. See MoveToAmend. Org for more on this.

The Adoption of Electric Vehicles

A friend sent me an article by Neal Asbury on the adoption of electric vehicles for my comment. Like so many other things, there are two sides to the "EV" coin, but, as I told him, this piece was written for the lowest possible common denominator of audiences; I've seen more credible journalism in the National Enquirer covering alien abductions and three-headed babies. (Of course, the 45% of 2GreenEnergy readers outside the U.S. have no idea what I'm talking about here. Sorry. You have to see it to believe it.)

Just how on-target is the author's point, i.e., that government is forcing new, impractical forms of transportation down your throat? It's true that the government is playing a role in the transportation of the 21st Century, just like it did in the 20th, when it subsidized domestic oil exploration, built the roads and highways, and consistently deployed the military to maintain access to oil from foreign sources.

Also, let's not forget that this subject of transportation does not end with cars and trucks. Does the article's author know where the radar systems came from that have virtually eliminated fatal accidents on our commercial airlines? Could he have told you that the annual risk of being killed in a plane crash for the average American is about 1 in 11 million, as compared with the annual risk of being killed in a motor vehicle crash, which is more than 2000 times greater (approximately 1 in 5000). And speaking of the government and the automobile industry, does he know the source of every single safety advancement, from seat belts in 1961 to anti-lock braking, air-bags, and the many dozens of other technologies that save more lives each year (though each one was fought tooth-and-nail by the auto industry itself)?

Expanding the concept of transportation one step further, has Mr. Asbury asked himself how we put a man on the moon and began to explore vast regions of the universe? Taking the subject of transportation back down to Earth, has he considered where the Internet came from, that provides us our real-time traffic maps, our roadside assistance, the backbone for deployment of emergency medical services, not to mention the hundreds of other benefits we count on every day?

Government's encouraging progressive concepts in transportation — pragmatic concepts that have proven themselves thousands of times over for their effectiveness in protecting your life and providing the safekeeping of your loved ones — versus "forcing something down your throat" are grossly different concepts.

At the risk of appearing rude, Mr. Asbury's talents would be better plied in covering Elvis sightings.

News From the Renewable Energy Policy Forum

As I was leaving the Renewable Energy Policy Forum on Capitol Hill earlier this year, ready for a brisk walk a few blocks north to Union Station, I ran into a fellow who caught my eye and said, "Not a lot of new news there, was there?"

"Oh, I thought there were some interesting insights," I replied.

"Can you name one?" he insisted.

For what it may be worth, here are a few abbreviated takeaways:

• The story of renewable energy recently is a mixture of triumph and disaster. Renewables in the U.S., even excluding hydroelectricity, doubled from 2008 – 2012, making good on one of Obama's earliest campaign promises, and, some would say, established clean energy as a bona-fide industry. And that's just the start: Prices are coming down, the subsidies for fossil fuels may be winding to an end, public acceptance of and demand for solar and wind is large and getting larger, and the laws that have hampered capital formation for clean energy projects are likely to change, creating a more level playing field. But then there's the "disaster" side of the story: natural gas prices have made it hard for renewables to compete, and our total demand for electricity is down, due to the recession, efficiency, and the retirement of the dirtiest of the coal plants. Many of the states' RPSs (renewable portfolio standards) have been met, thus removing further impetus from this once-powerful driving force. And let's note that the oil companies still essentially own our Congress; the will of the people is continually frustrated by this form of corruption.

- A presenter from venture capital firm DBL says that her organization likes to ask, "What would Jefferson do?" She noted that government has played a fundamental role in the development of our country, from the land grants encouraging the pioneers, to the creation of the railroads, to the formation of the oil industry, the highway system, and, more recently, nuclear power. The concept that government should, once again, push our society in a positive direction doesn't seem at all out of line, given this extremely consistent history. She went into detail about the success of a great many of the cleantech companies that have been the beneficiaries of federal loan guarantees and government grants under ARPA-e. Having said this, the portion of the federal budget that is allocated to entitlements (vs. discretionary spending) has risen from 30% in 1970 to 70% today, thus there are significant limits to what can be accomplished here and now.

- The sad fact that the U.S. has no energy policy (and thus serves up wild, unpredictable swings in critically important legislation) produces gross inefficiencies in our country's attempts to develop renewable energy.

• Unsurprisingly, the idea of "turf" is a big deal in Washington. Want to do something in hydrokinetics? That means that you'll be dealing with most or all of the 40 – 50 agencies that deal in water rights, fish and game, recreation, shipping, maritime commerce, etc. Then, on top of the government, brace yourself to deal with the lawyers who represent private interests that may be threatened by your proposal.

• Contrast all of this with Germany and the other countries that have streamlined these processes. Want to do a solar project in Germany? It doesn't take years and huge sums of speculative development capital; it takes just a matter of days. Banks understand the subject and eagerly lend money to support it. Perhaps more importantly, government understands the subject as well, and has made the process extremely quick, straightforward, and inexpensive. There's so much solar in Germany (12 times more per capita than the U.S.) that lawyers dealing in the area are almost non-existent. Because of all this, solar is about half the price per installed watt – not based on the cost of modules, but on the cost of project bureaucracy.

• The biofuels people were out in force at the Policy Forum; of the 20-or-so speakers, at least four or five were banging the drum for bio-ethanol / bio-diesel. They claim, and I'm sure it's true, that they are treated quite unfairly by the petroleum industry, who has fought the RFSs (renewable fuel standards) tooth and nail. But I found it strange that the term "electric vehicle" was not mentioned once all day. I left with the impression I was afraid I'd have: this is Washington. It's not about fairness, openness, objectivity, or serving the people; it's about promoting the industry that represents your meal-ticket.

• Further evidence of this "ax to grind" phenomenon came when the lady from Lockheed Martin spoke. Every word out of her mouth was about marine energy, i.e., ocean current, and what a tragedy it is that the federal government will not subsidize this hugely profitable entity (celebrating its 100th birthday) in their development of ocean current hydrokinetics.

The simple truth is that I have unrealistic "Mr. Smith Goes to Washington" expectations from our government. I'm hoping that some person or some group will ask:

"What do the people of the United States really need? What can Washington do to help the majority of the people lead better, healthier lives? What can be done to restore America's greatness? What can our country do to lead the world in the 21st Century?"

Is there anyone in this town who thinks that way? Sure, but they're in the slim minority, and they're an endangered species, since, by definition, they're not taking much-needed campaign contributions in exchange for favors. Let's put it this way: for every one of them there are at least 100 people on a mission to forward a specific, money-driven agenda of some sort of special interest.

So, was there any "new news" there? I'll let you be the judge.

The Adoption Curve for Electric Vehicles

We frequently come across articles that present specious logic associated with electric vehicle adoption. In particular, the greening of conventional vehicles militates away from, not towards, the adoption of EVs; the payback in fuel consumption for an EV is far more attractive when the car one's replacing gets 25 MPG, rather than 60 MPG.

Having said this, I do see a day when the case for electric transportation becomes overwhelming, both for the individual and for society. Imagine, if you will, a time in which:

• EV range issues will have all but disappeared, i.e., ranges of 300 miles have been achieved at a reasonable cost. Other costs have fallen as well, due to economies of scale and advancements in technology.

• In addition to home and the workplace, "opportunity charging" locations are springing up, including fast-charging locations. (Note that this is already happening; see Tesla's project for travelers.)

• Society decides to "internalize the externalities" of gasoline, i.e., create a landscape in which we pay the true costs of fossil fuel consumption, including the damage to our environment and our lungs — and, dare I say it? — the use of our military to provide access to oil. (While this is anything but a slam dunk, it's possible; in fact, I argue that it's becoming more probable every year as more people start to see the horrific impact that fossil fuels are creating on our planet.)

• EVs become fairly innocuous in terms of their impact on the grid. This is not too hard to imagine, by the way. Coal-fired power plants are steadily being retired, starting with the oldest, i.e., dirtiest. I'd hate to see it, but if we really are building more nuclear power plants, that's 24/7 base-load, and, since EVs are largely charged at night with off-peak power, this is helpful to the cause. EVs also create a home for wind energy that is currently curtailed or sold at negative prices.

• The presence of EVs encourages all manner of other futuristic concepts: distributed generation (e.g., mid-sized wind), as well as smart-grid, including vehicle-to-grid (V2G) technology.

Again, this is "down the road" thinking, but not too hard to envision.

P.S. (written a few days later)

When I wrote above that, "The greening of conventional vehicles militates away from, not towards, the adoption of EVs; the payback in fuel consumption for an EV is far more attractive when the car one's replacing gets

25 MPG, rather than 60 MPG," I knew I'd horrify some people. Here's one such response:

I was horrified to see a kind of fundamentalism that exists on the issue of EV and smart grid developments, that people would not appreciate upgrading from conventional vehicles to hybrids; or the attraction or price advantage or payback would not be justified for going towards EVs. ... It almost angered me to the point of thinking of jumping out of the window. Such infighting cannot take us to a sustainable future.

While I understand such thinking, I suggest a sense of pragmatism in the face of the real world. I don't see this as infighting; I see it as reality. "Solutions" to our problems that are not economically attractive are not solutions at all.

Even if you view that as bad news, here's the good news: cleantech in all its forms: renewable energy, electric transportation, energy efficiency and smart-grid, etc. – all carry with them extremely powerful economic arguments that are becoming clearer and more forceful with every passing day. Yes, we have some political hurdles we need to cross, but the world is very close to jumping into cleantech with both feet.

If you remain unconvinced, here's a tiny proof-point. When I attend the Renewable Energy Finance Forum twice a year, I don't see a few tiny venture capitalists wringing their hands about the prospect of making investments of a few million dollars. Rather, I see Morgan Stanley, Credit Suisse, Deutsche Bank, Bank of America and CitiGroup — real financial institutions, as well as some very, very well-funded venture capitalists and equity investors that are literally putting up billions of dollars. This really *is* happening.

There. I hope I've talked my reader down from his window ledge.

"Climate One" Produces Fabulous Conference on Consumer Adoption of Green Products

I attended a terrific conference earlier this year produced by "Climate One" at The Commonwealth Club in San Francisco. For those who may not be aware of this incredible organization, formed 88 years ago, "The Commonwealth Club of California is the nation's oldest and largest public affairs forum."

What does that mean? Think: "TED Talks" – "ideas worth sharing" – talks LONG before there was an Internet by which they could be shared so easily. And think: no political spin, as hard to believe as that may be to comprehend. There are democrats and republicans, liberals and conservatives, scientists and politicians; there is UN Secretary General Kofi Annan, Archbishop Desmond Tutu, and Vice President Dan Quayle. Just important ideas, good reasoning, and great communication skills (OK, I know what you're thinking about Dan Quayle, but you didn't hear me say it, did you?)

The conference I attended Friday covered the consumer acceptance of environmentalism – the adoption of eco-friendly products, and the rush of product marketers to add "green" messaging, whether honest or not, to their brands. I hope readers will be interested in the notes I took from the conference, which I reproduce here; I've italicized the comments I added on later.

• There is an increase in awareness and demand for green products, and thus a proliferation of such products. Green products are the fastest growing sectors of cars (hybrids and EVs), food (organics), cleansers, etc. Though they are the fastest growth segments, we're talking about growth that is starting from very small bases; note that going from 1% of a market to 2% in a certain product category is 100% growth. *True. But from my perspective, all movements start from a small group of early adopters; there is reason to believe that we're seeing the formation of a huge event in human history.*

• Sustainability needs to be more than a corporation's going through the motions for the PR value; it needs to be elevated to the very top of the organization and become a part of the culture/DNA. *I've heard this a million times, but I'm skeptical. Sustainability, in my mind, ultimately means leading this massive population away from blind consumerism into a lower carbon way of life. Companies want people to buy more of their stuff, and most of the world's successful corporate entities will come to ruin if we truly head in a sustainable direction. Coca Cola and McDonalds want more people drinking more sodas and eating more Chicken McNuggets, regardless of how many trees they're planting to distract people from their atrocities. Toyota had the PR coup of the last 50 years with the Prius, but they sell far more Sequoias (13/18 MPG) and the dozens of other planet-killing members of the fleet. The average electric drill is used for 9 minutes from the time it's taken from Home Depot's shelf to the time it finds its way into a landfill. As gross as that is, it causes heartburn neither to the people who built it, nor to Home Depot who sold it.*

• The U.S. Federal Trade Commission (FTC) has, after 18 years, finally come out with an edict banning the most egregious forms of green-washing, i.e., lying about the eco-friendly virtues of your product. This was hailed as a disappointment, as, in those 18 years, the world has become far more sophisticated in deceiving the public, and the FTC has barely caught up to where green-washing was ca. 1990. *I have to admit that policing this would not be easy for any governmental agency, regardless of how sharp and honest. Consider that the Lexus SUV hybrid gets worse gas mileage than the non-hybrid version. Of course, this didn't slow Lexus down from promoting it to its eco-conscious buyers — but what can government do about that? The world is powerless to deal with issues like these if people are really that dishonest.*

• Consumer messaging is complicated by the fact that it needs to be extremely simple; consumers in supermarkets, for example, even those who make any effort at all to determine the eco-characteristics of the products they're buying, spend just a few seconds making their decisions. It's unrealistic to expect any reasonable percentage of consumers to sort through the relative vices and virtues of, say, a rain jacket, that may have been manufactured with well-treated labor, sent to the U.S. with highly fuel-efficient ships, but may have been treated with a toxic chemical to achieve waterproofing. No one can be expected to absorb and make sense of all this information. Thus, eco-conscious people increasingly tend to trust their peers – perhaps out of necessity; the level of trust that we place in corporations is the lowest ever, in the entire history of measuring this statistic. *Yes, it's true that* we *don't trust them, and yes, we do have access to a wider set of peer-based information on which to make our choices.*

• Corporations that try to work around this with clever attempts to harness the power of social media often wish they hadn't. A good example of this is what is referred to as "bashtags" (as opposed to hashtags). When McDonalds thought it could generate good PR by taking advantage of the vast audiences on Twitter and FaceBook and asked people to share their "first McDonalds experiences," I'm sure they thought it would elicit stories of mommies bringing their little ones for burgers and French fries after their first little league games. Instead, the comments, which went mega-viral, were things like: "I heard about the rat feces in BigMacs," or "I learned that McDonalds is the biggest single source of deforestation on the planet."

Something to watch out for, to be sure. The campaign "Chevron Does" met a similar fate. The intention of the campaign, obviously, was to highlight Chevron's many claims to humanitarian activities, e.g., "Who invests in renewables? Chevron does." Within minutes, however, there were swarms of suggestions like this: "Who ruins the Ecuadorian rainforest?" and "Who profits at the expense of our health?" *Ouch.*

• Sustainable products, to be successful, need to be high-quality and well-priced. *I'm sure this is true, but here's the challenge: There is a reason that unsustainable products are cheap: there are externalities that are being passed on to innocent bystanders. This hits close to home here at 2GreenEnergy, as coal is the cheapest form of energy. Of course it is! No one is forcing the coal companies, nor the consumers of the electricity from coal-fired power plants, to pay for any of the damage they're causing to our lungs or our ecosystems. Clean energy solutions are shunned because they cost more; obviously they cost more than solutions that pass the majority of their costs on to the unwitting customers (and their grandchildren).*

• Zipcar, the car-sharing success story that Avis bought this week for $500 million, is an example of what is called the new "collaborative economy." *I'd like to think there is some truth there; I guess we'll see. Apparently, Bill Ford is contemplating morphing his company from the traditional car business of the 20th Century to the "mobility" business of the 21st. I.e., he's claiming that he's thinking beyond the concept of car ownership for as many people aged 16 – 96 as can afford it. Again, I'm skeptical. When the mobility paradigm changes, and*

I think it's in the process of doing exactly that, I'm not sure you'll want to be holding onto Ford stock.

• *So far, this has all been fairly good news for the human race. The bad news is that most people in the U.S. really don't care about any of this at all.* In surveys, 75% say that we would pay a bit more for an eco-friendly product, but in practice, only 1% – 3% actually do. Apparently, we suffer from what's called "green fatigue." The "green" story is old and getting older; evidently, people are tired of hearing it. *I can understand that. On Wednesday afternoon, our media tells us about Lindsay Lohan's drug relapse; the following morning, we've all heard too much about it, and so it's on to something else. This cultural ADHD is a total mismatch for the problem here. People say that recovering alcoholics are in recovery until the day they die. Fighting a lifelong battle of anything, especially slow-rolling things like the destruction of our environment or of our social fabric caused by our addiction to fossil fuels, is the LAST thing the average American wants to do. As long as there's plenty of Budweiser and Doritos at the Walmart, and a Toyota LandCruiser to pick them up in, the typical American is comfortably numb.*

• *Looking at the problem differently, we environmentalists often have trouble communicating our messages, as our culture responds to visual images.* We had minute-to-minute TV coverage of the BP oil spill and the disaster at Fukushima, as those events lent themselves to imagery. But as we speak, tens of millions of children's lungs are being ruined by the effects of coal. There's no way to tell that in an exciting, newsworthy story with real-time televised images.

Great event. I'm so glad I took the time.

Renewable Energy "Industry" Achievements for 2012

The people at ACORE, the American Council on Renewable Energy totally "get it."

Below is a reprint of part of their piece, "Renewable Energy Industry Achievements for 2012." In case it's not obvious, the operative word here is "industry." ACORE understands that the real obstacle that clean energy faces is its status as a bona fide sector of our economy, i.e., its standing against an *industry*.

After all, it faces vigorous opposition from the fossil fuel boys, and if there ever were an industry, that would be it. Perhaps the term "empire" is more apt.

Here's a group that, by all accounts, owns Congress. If you think I'm exaggerating, simply note that the brave senators who voted last March to repeal the subsidies for the oil companies received one-fifth the campaign contributions per capita from Big Oil than those who ever voted to continue them.

As I've reported previously, ex-Pennsylvania governor Ed Rendell speaks very precisely as to why clean energy is having a tough time. According to Rendell:

"There are too many special interests arrayed against it. Over 90% of Democratic voters are in favor of Congress passing legislation that prioritizes clean energy. In fact, over 75% of Republican voters are in favor of the exact same thing. Clearly, the will of the American people is being frustrated by special interests.

"Together, we can do this, but we can't do it inside the Beltway. The lobbyists are raising campaign money for our senators and representatives in Washington seven days a week. It never stops. It never stops. There are fund-raisers happening literally every night. If change is going to happen, it needs to take place in hometown America. Your leaders have to hear it from you."

Going back to ACORE, here are some of the renewable energy industry's accomplishments last year:

Last year was marked by great successes for our renewable energy industry. In the first 10 months of 2012, 46.22% of new electrical generating capacity brought online was from renewable energy.

- *The solar energy industry grew at an incredible rate of 13.2% and outpaced the growth of the economy.*
- *Wind power added 4GW in the first 9 months of 2012.*
- *Globally, geothermal supplies 11GW of electricity and the U.S. is still the global geothermal leader.*
- *Hydroelectric power generation increased by an annual average of 3% and new developments in tidal energy show the potential for commercial-scale tidal energy in America.*
- *Biofuels aided American families at the pump and the EPA's decision to deny a partial waiver for the Renewable Fuel*

Standard continued to help American consumers, our military, and the biofuels industry.

- *Electric vehicles experienced record breaking sales in the last few months of 2012.*
- *The waste heat power sector now has the potential to install 10GW, enough to power 10 million American homes.*
- *This progress did not happen by accident; it happened because of your hard work and dedication to bringing renewable energy to scale in America. It happened because we all believe clean, renewable energy, strongly coupled with energy efficiency, is fundamental to fueling a more prosperous, sustainable future for our country.*

My heartiest congratulations to the entire industry, and to everyone who honestly cares about the quality of life we leave for our descendants.

21st Century Planet Earth — Fabulously Uninterested in Confronting Its Challenges

December 15th was the birthday of Nero, the Roman emperor who is said to have fiddled while the city burned in 64 CE. It's not clear whether the fire was a case of arson, or just an accident, which wasn't unusual in the day.

In any case, whenever I come across a reference to "fiddling while Rome burns," I'm compelled to think of the parallels to our civilization today, and how we as a society are so strangely uninterested in and disconnected from the dangers we face in terms of wrecking our home planet. Collectively, we carry on with business as usual, consuming an ever-growing amount of fossil-fuel energy and other resources, shrugging our shoulders at the growing damage that derives from our irresponsible ways.

For a moment, think about what it will take to change course here, and begin to make a few sacrifices that would be required in order to put us on a path to sustainable resource stewardship. Then ask: Are we cut out for this? Do we have this in our DNA? When we look at the behavior of the various nation-states and the most prominent people and potent forces within them, it sure looks like we're a million miles from even the most basic conditions under which we can talk meaningfully about changing our course.

I applaud folks like Rajendra Pachauri. Here's a brilliant and supremely well-educated guy, boldly trying to lead the Conference of Parties meetings and navigate them in a productive direction. But the odds he faces are long in the extreme.

At the risk of oversimplification, the challenges seem to come in two broad forms:

1) What could loosely be called "insanity." We live in a world in which North Korea just launched a missile into space, Iran is desperate to enrich uranium, the U.S. defense budget is larger than that of the 17 next smaller countries in military spending combined, and 1.5 billion people can't get a clean glass of drinking water.

2) Entrenched interests whose exclusive concerns are profit. Dr. Pachauri is a scientist with not one, but two PhDs, but he's not an autocrat; all he can do is present facts and hope that they inspire action. How likely is it that he's going to be able to change the behavior of nations whose true power structure is in fact rooted in the oil companies, hell-bent as they are on sucking the last molecule of crude oil out of the ground, when these companies employ more lobbyists than any other industry in the known universe? Can scientists take on an industry that earns a profit of $375 million — *each day?*

I'm not saying it's impossible for us to put that fiddle and bow back in their case, and confront the truth. But, as a challenge, it's a *beast*.

Building Energy Policy on Facts, Not Dollars

It's common for environmentalists to take knee-jerk positions, rooted in an incomplete and self-serving view of the relevant science, and these positions can ultimately do more harm to the environment than good. Sad but true.

Can thoughtful and fair-minded people see a case for genetically modified food? Can nuclear energy and shale gas play a role in mitigating the planet-wrecking horrors of coal? I believe the answer to both questions is Yes. As Glenn Doty of Doty WindFuels likes to say, "This is a marathon. Sprinters will not cross the finish line."

So yes, let's admit that a rigid, "don't confuse me with facts" position is childish and indefensible, regardless of the issue and which side of it you believe you're on.

But my concern is that, as a society, our decision-making processes aren't really based on facts at all; they're based on money.

Take GMOs as an example. Are they dangerous? I don't know. They certainly don't seem to be, based on incredibly vast amounts of carefully collected data. My concern about GMOs isn't that they're dangerous; it's that neither you, nor I, nor our elected representatives had even a tiny peep of a voice in the decision to move GMOs into the marketplace. That decision was made by Monsanto and a few other mega-corporations, whose unfathomable power steamrolls the process that we might hope would regulate a decision of this magnitude.

This, of course, gets us back to the U.S. Supreme Court decision "Citizens United" and how, until it's overturned, the "corporatocracy" described in the Monsanto/GMOs case will remain pervasive here in the United States. Everything we do that has any real monetary value, whether it's the consumption of food, energy, transportation, healthcare/pharmaceuticals, etc. is mandated by a few people at the top of the corporate world who make the decisions as to how we live and, often, how we die. We get excited about the decisions we make each two years in our election cycles, but we're kidding ourselves if we believe we have any meaningful participation in our government.

Thus, it's probably a good time for another plug for "MoveToAmend.org" and Bernie Sanders' initiative: "Saving American Democracy." Someone wrote the other day that the subsidies we hand out to solar are disproportionately large compared to those for wind, given their capability to mitigate greenhouse gas emissions, to which I replied: *What's really required here is fair-mindedness. As long as decisions are being made to favor a concept purely on the basis of how much clout it has, we're doomed.*

Discussion on Electric Transportation Is Part of a Larger Topic

When asked what 2GreenEnergy is, I normally talk about what we're doing to forward the cause of clean energy, but I quickly add that we're about sustainability more generally. I want to ensure that we play a role in any aspect of cleantech applied to transportation, agriculture, or any other discipline that affects humankind's long-term viability.

Here's a good example — our ongoing conversation on electric transportation, in which frequent commenter 2GreenEnergy Glenn Doty wrote:

Since the externalities of fossil fuels are not priced into the market equation, there is literally zero chance that EV's will make a dent in the market. ... Costs matter .... Upgrading the American transportation system to EV's, then upgrading the electricity grid to distribute sufficient energy from the wind-rich plains to the entire U.S., then upgrading storage across America to resolve the intermittency issues with wind and solar, then installing rapid-charging infrastructure across America, then upgrading the last-mile energy capacity of the grid to accommodate widespread at-home charging...

All told the price tag for the above will be in excess of 15 trillion dollars. That's not a typo, and it's not an exaggeration. This is essentially a dead cost to the consumer so they can drive one type of car rather than driving another type of car...

Glenn has a terrific grasp on math, but I believe that the idea that the consumer needs to absorb his share of $15 trillion so that he can "drive one type of car versus another" misses a few points. First, let's talk about that $15 trillion:

• This is money that will be spent over a considerable period of time.

• The infrastructure by which we deliver electricity to our populace is rapidly falling apart and needs to be upgraded anyway.

• This upgrade is especially necessary if we expect to make any real progress in ratcheting up our use of renewable energy to drive down our use of fossil fuels. Let's review the imperative here. The vast majority of climate scientists tell us that our dependence on coal, oil, and natural gas is causing climate change, ocean acidification, ever-increasing rates of lung disease, loss of biodiversity, and dozens of other disasters to our personal health, and to that of our natural environment. The cost of these, if they can be measured at all, is stratospheric. Most scientists tell us that our civilization will collapse if we don't take aggressive action in this arena on a planetary level. But renewable resources tend to exist in specific geographical regions, making energy transmission an important objective.

• The efficiencies brought about by upgrading our grid are enormous. For example, the concept that we call "smart-grid" removes huge chunks of cost, mostly by reducing energy consumption. Granted, these cost reductions are measured in tens of billions, not trillions of dollars.

• Smart-grid also changes the condition that a large percentage of the cost of delivering electricity to consumers and businesses is built around approximately 100 hours a year, when demand is at its peak.

• The advent of electric transportation, since it's normally recharged at night when demand for power is low and wind energy capacity is high, has the potential to help us integrate more renewables onto the grid.

• Not that professional economists have the same standing in our scientific world that they enjoyed a few years ago, but most of them have extremely convincing arguments to the effect that putting the required millions of people to work on these endeavors will rejuvenate the economy and expand the tax base. They have some fairly precise numbers that they project here, and I find them compelling. Those who read my second book (<u>Is Renewable Really Doable?</u>) will recall my interview with Dr. Robert Pollin, Co-Director of the Political Economy Research Institute, on the subject.

• Such an effort would also help to rebuild the battered middle class, and re-establish the U.S. as the superpower it was in the 20th Century. It's clear that energy is emerging as the single most important industry in the 21st Century, and Americans are at a loss to explain why the country they love is sitting on the sidelines while dozens of countries in South America, Europe and Asia, most notably China, are leading the world. Obviously, it's impossible to assign a dollar value to this benefit, but no one doubts that it is of immense importance at a national level.

• A better grid with stronger resiliency will reduce the frequency and duration of power outages that, due to extreme weather events, are increasingly common, often with lethal effects.

• This planet may not be running out of oil, but it's certainly running out of *cheap* oil. The costs, both financially and ecologically, of extracting and refining oil from tar sands and shale, are far greater than most people believe.

• Since its onset a century ago, oil dependence has caused global hostilities. It's hard to know how to put a price on a young man, who, serving in the Middle East, comes back in a bag. Here's an excerpt from my 2010 interview with James Woolsey, ex-U.S. CIA Director:

Oil, like gold before it, has the effect that Paul Collier at Oxford, and Tom Friedman cite sometimes called the "oil curse." Generally it's just that an autocratic state, when it depends for a huge share of its income on a commodity that has a lot of economic rent attached to it, that rent accrues to the central power of the state essentially. So you tend not to have representative institutions like legislatures and you tend to have a much more difficult time getting out of an autocratic structure than with a broad-based economy.

If you look at evolution, the examples I usually use are Taiwan and South Korea. They were tough dictatorships, but as they prospered and built up a middle class — and this happened to them a lot faster than it happened in Europe in the medieval and early modern times — it was a similar phenomenon. The middle class builds up, it's diversified, it starts wanting economic liberties and that transmogrifies after a while into political liberties and it tends to gravitate toward freer institutions. That tends not to happen when you've got a lot of economic rent associated with a commodity that you're heavily dependent on. Read Larry Diamond's book if you haven't already. If you look at the

22 countries that count on two-thirds or more of their national income from oil — it's fair to say all 22 of those countries are autocratic kingdoms or dictatorships.

And I haven't compared that list with Freedom House's list of the 40 basically – those that Freedom House calls "Not Free." There are about 120 democracies in the world, not perfect, but nonetheless regular elections and another 20 countries like Bahrain that are reasonably well and decently governed, even though not democratically so. And then you've got 40 really bad guys. And I'm pretty sure that list of 22 in Larry Diamond's book is virtually all from the list of 40 bad guys — or "Not Free," in Freedom House's terms.

Finally, as Glenn Doty correctly points out, "Costs matter." He's also right that the externalities of fossil fuels are not priced into the market equation. But how much longer can this possibly be the case? How feasible is it that humankind will fail to restrain itself, and then go forward and poison itself into extinction? As strong as the oil companies are, and as powerfully as their tentacles constrict our political discourse and frustrate our attempts to cut ourselves free from them, it's only a matter of time until the world wakes up to the truth.

In his farewell address in 1961, Dwight D. Eisenhower, warned us:

"You and I, and our government must avoid the impulse to live only for today, plundering, for our own ease and convenience, the precious resources of tomorrow."

I was only six years old at the time, quite unaware of the insight that Ike had bestowed upon us. Of course, he wasn't the first. 150 years earlier, Thomas Jefferson wrote:

"It is incumbent on every generation to pay its own debts as it goes."

In short, we have a huge problem that we need to confront. For us to continue to party like rock stars on cheap energy (whose true cost is many times higher than what we're paying for it) is simply wrong. Let's pony up the resources to address this situation now, right now, given that, if we ignore this opportunity, the costs will only increase over time.

Dealing with Population Growth and Issues of Sustainability

Recently, the World Future Society published an article on population growth and sustainability that presents the idea that we're consuming resources at an unsustainable rate, and that rate will only increase for the foreseeable future. Right now, we'd need an Earth about 1.6 times the size of our own to provide the resources that our population consumes annually. And as each year passes, the number becomes larger.

Put another way, there was a moment during a day this past September at which humankind had consumed all the resources that would be naturally renewed in the entire year of 2012. Soon, that moment will be in August; a decade or so later, July.

So what to do?

Each of the cleantech business plans featured on 2GreenEnergy.com represents my main answers. They're here precisely because I believe they contain some level of breakthrough technology that, if implemented, will move the needle associated with humankind's potential to sustain itself.

Let's take an example: the cutting-edge concept of aeroponics. If I were speaking to a group here, instead of writing, I'd ask for a show of hands. How many of you believe that our current approach to agriculture, with its ever-increasing quantity and potency of chemical fertilizers and poisons, can sustainably support a population of seven billion, on a direct climb to 10 billion? I wouldn't embarrass the few who had raised their hands, but I would present a few facts that show how completely unsupportable this notion is. I would ask if anyone knows the average distance that the food in our grocery stores traveled by diesel truck to get there. I would congratulate those who know the answer (1200 miles). I would ask if that model sounds sustainable. I'm confident that, within a few minutes, I'd have a large percentage of the audience converted to the concepts of aeroponics, bioaeroponics, and aquaponics.

Of the other "big ideas" that I support in terms of sustainability, the most important are probably those that drive education. The most frightening challenges we face in dealing with our skyrocketing world population are rooted in ignorance. Uneducated people tend to have more children and fail to educate them, thus perpetuating the cycle of poverty and runaway population growth. That's why I think the business plans that will develop rural off-grid/ microgrid electrification, microwind and mid-sized wind, are so important; they will foster education in regions of the world in which it's currently extremely rare.

Others address climate change, like this unique approach to carbon-neutral synthetic fuels.

Again, each of these cleantech business opportunities is there for a reason. If you know anyone who could potentially play a role as a partner or investor, please let me know.

The Tough Realities of Wind and Solar

Earlier this year, frequent 2GreenEnergy commenter Steven Andrews wrote:

Wind and solar have the best characteristics plus on top of that use no water resources. Why are these systems so under attack by fossil fuel burners? Gandhi said: first they don't pay attention, then they deny, then they attack, lastly, we win. Wind and solar don't consume oil, gas or coal, they don't leave toxic chemicals, or radioactive materials behind, so they are cheaper and better.

Better, yes — in several senses; cheaper, no. Wind is comparable in cost to coal, but its intermittence creates an issue (and thus expense) for those wishing to integrate it into the grid-mix in a big way. If we don't put a cost on coal that covers the damage to our health and our environment, we'll be burning coal for a very long time.

Having said that, we *will* get to widespread renewable energy, as you point out with your wonderful reference to Gandhi. The only questions are: a) how much damage we will have done to the ecosystem, and b) who's going to make a buck in the process.

Big Energy, the people who got obscenely rich in the 20th Century, doesn't care about the damage, and want to make sure their monopolies continue intact. Are there forces with sufficient power to derail them? We'll see. According to Victor Hugo, "There is nothing more powerful than an idea whose time has come." Obviously, I'm one of the many millions trying to make sure that clean energy is indeed, an idea whose time has come.

Does humankind possess the survival instinct and raw guts required to defy Big Energy's ever-increasing assault on our precious, dying planet? Again, we'll see. When I speak with my mother about this, we often lament that neither of us will be around in 2100 to see what happens as a result of what I see as the selfish and short-sighted policies of 2012.

Climate Change and Extreme Weather Events

The public relations machine that works so hard to generate doubt about climate change and the extreme weather events it causes must have its hands full at this point.

2012 was a year in which the United States had no winter, a March with most of the country above 80 degrees F, floods in its three largest rivers, a horrific drought all summer, and now Hurricane Sandy, with its loss of life, tens of billions of dollars in damages, and incredible expense and inconvenience to many millions of people. I'm not sure how you get people not to notice something as obvious.

The fossil fuel industry is by far the most profitable in the history of humankind, and thus can afford the very best and loudest of voices to protect its interests. They've spent an utter fortune obscuring the facts concerning climate change, and they've been fantastically successful in confusing a huge segment of the American public. But at a certain point, they'll find themselves unable to fend off a rising tide of public opinion that screams to its government, "Do something! Help us! You're suppose to be our *leaders!*"

Keep in mind that we as a species are not powerless to deal with environmental issues on a global scale, as evidenced by the effectiveness with which we repaired the hole in the ozone layer when we discovered it in 1985. Granted, that fix (a ban on chlorofluorocarbons (CFCs) was far easier than dealing with the current crisis. Having said that, the moment we agree to price in the externalities of the energy we produce and consume, e.g., the costs to human health and the natural environment, you'll see an instantaneous explosion in the development of clean energy and efficiency solutions, along with stunning levels of conservation.

We could even emulate the Chinese, who, unlike the U.S., have a 21st Century Energy Plan, and are busily making it happen. They have decided that it's tantamount to national suicide to sit around and act like us Americans, bickering about renewable energy, or, in the case of the presidential debates, pretending it doesn't exist and ignoring it altogether. The Chinese are hard at work, implementing all kinds of cleantech wonders: ultra-high voltage electricity transmission, electric transportation, smart-grid, and dozens of other cutting-edge solutions that will propel them swiftly and irreversibly into the position of world economic leader.

But I predict that Americans will not tolerate this indefinitely. Many of us can observe things for ourselves, and ignore the garbage we're being told. The people telling us to "pay no attention to that man behind the curtain" are rapidly running just as low on credibility as did the Wonderful Wizard of Oz.

Cleantech Investors Meet Entrepreneurs at 2GreenEnergy

"We should look inward and think about the meaning of our life and its purposes, lest we do it in 20 or 30 years and it's too late." – Robert Coles

As a professional, a husband, a father, and a citizen of the world, I think about this a great deal. In terms of the "citizen of the world" category, by the time I check out of here, I'm hoping to be able to say that I was quite effective at identifying really good concepts in cleantech, and moving them forward.

That, of course, is the thinking behind the 2GreenEnergy website.

My job, and I take it seriously, is to reach out to the investment community, into which I'm maneuvering myself further each day, and say: "Here's the result of my efforts. Here are ideas that have huge potential implications, the science of which is not smoke and mirrors, the teams behind which have proven that they have what it takes to be successful. Do you need to do your own due diligence? Of course. Have I added value in vetting 1200+ business concepts over the last 3+ years? I'd like to think so."

If you haven't seen it recently, I urge you to visit our so-called "Investor's Page" and scroll down slowly and thoughtfully. As always, if you know of anyone interested in playing a role, please let me know.

Provocative Questions on Energy from the U.K.

Earlier this year I did an interview for a young man in England, James Alcock, who writes for a consumer-oriented website called TheGreenAge.com. In a period of about half an hour, I answered various questions ranging from electric transportation to fossil fuels to energy storage and renewables. It's a delight to get a chance to answer good questions on the subject, as I always leave such discussions with some realizations and new viewpoints.

Here's something I found thought-provoking: James asked me if I recalled the precise moment that I decided to get involved in renewable energy, and the thoughts I was having at the time. Eventually we got around to the challenges facing this whole evolution away from fossil fuels, which I summarize as follows:

The world has very little appetite for short-term pain, even with the knowledge that its behavior is causing a global catastrophe in the long-term. People are unwilling to make sacrifices for the public good, and leaders are elected by promising the people exactly what they want: in this case, cheap energy. The fossil fuel companies, the most profitable industry in the history of humankind, exacerbate all this by dumping huge sums of money into public relations campaigns that promote an elaborate web of half-truths and outright lies.

The 58 million people who watched the 2012 presidential debates in the U.S. saw all this hit them full blast. The leader for roughly half of Americans openly ridicules the concept of clean energy and climate change mitigation, and the leader for the other half essentially avoids the subject.

But our antipathy towards a sustainable energy policy won't last forever; in fact, it can't. As I told James, "The 21st Century will see a wholesale change in the way we think about our responsibilities to one another, how we behave as citizens of the world. This is emphatically not happening in the U.S. right now, but I predict it will at a certain point. Selfishness has been the hallmark of human civilization since it took root about 11,000 years ago, but ignoring the needs of others, coupled with the products of the Industrial Revolution, has brought us to the edge of an abyss – face to face with the catastrophic collapse of our ecosystem. I predict that humankind will realize what's happening and make the changes it needs to before it goes over the edge. I hope I'm still on the planet to witness it, because it will be a thing of profound beauty."

An Open Letter on Energy to Mr. Obama

After Wednesday night's debate, I became one of literally millions wishing to give U.S. President Barack Obama advice, and, for that reason, I certainly do not believe that what I'm about to write has any particular importance. Having said that, here's a brief "open letter."

Mr. President:

A great number of us who viewed your debate on October 3rd, 2012 were astonished that you were either unwilling or unable to defend yourself vis-à-vis America's energy policy, and thus I offer a few "talking points" for your consideration on the subject:

Every major nation on Earth is making significant investments in energy efficiency and clean energy, as this is the industry that will define leadership and success in the 21st Century. Here, I mean "success" in every meaningful sense of the word: technological, economic, social, moral, and ultimately, military. Supporting a measured but steady migration away from fossil fuels is a clear winner across every single dimension. If in the context of the upcoming debates you choose to stress economic success over the others, I can understand that given the circumstances. But any way one looks at it, embracing clean energy is the single most pragmatic thing that anyone who loves this great country can do.

The United States needs to make investments from the public sector, while encouraging them in the private sector. To the degree we do not, we are actively rushing towards a new status as a kind of underling in the international community. These are the moments, unfolding over the next few years, in which we will either stand up and play a key role, as we did in every important industry throughout the 20th Century, or fall gradually into irrelevance. Yes, we could be crushed under the weight of the Republican demagogues in their last-ditch desperation to win in this election, and their subsequent eradication of every bit of progress this country has made over the last 50 years in terms of environmental protection and stewardship.

By the way, it's clear to almost everyone that these "demagogues" are the pawns of the large interests in traditional energy, and you know this very well. How candid you wish to be on this point is a matter of judgment. Whatever you do, you mustn't be tepid in your discussions of the subsidies for the oil companies. This is an evil and blatantly corrupt practice, and *we all know it* – even most Republicans. The Senators who voted to retain the subsidies last March received, per capita, five times the amount of campaign contributions from the oil companies as did those brave and honest enough to vote against them. If you're looking for the most obvious example of malfeasance in government, you needn't search any further. All you're earning by mincing words here is the bitter disappointment of your supporters who are counting on you for leadership, not some kind of cringing diplomacy. Whatever words you choose, you need to be proud and bold, not weak and defensive, about what we've done and where we need to go as a nation.

As you're well aware, the economists who have no axe to grind here point to an extremely bright future in terms of green jobs. Do not allow your opponent to get away with the idea that a sustainable approach to energy and transportation is a job-killer. The precise opposite is true; you know it, and you're more than able to defend this critical point.

Mr. President, it is my hope that before the next debate, you will take a few deep breaths and think back on this letter. Then go in and *nail this*. An anxious nation as well as people all over the world are watching, counting on you to come through as the leader you have the potential to be.

Best regards and good luck,
Craig Shields

Thoughts from the Renewable Energy Finance Forum

It was great to be in San Francisco for another extremely successful meeting of the Renewable Energy Finance Forum, an occasion that gave me the opportunity to chat with Dennis McGinn. Dennis is president of ACORE, the American Council on Renewable Energy, the organization putting on the show; he's doing a great job in forwarding the cause of a sustainable approach to the generation and consumption of energy.

Here are a few ideas that came to me in the course of the two-day event:

1) The mayor of Palo Alto told us that his city is doing a great job in terms of achieving its RPS (renewable portfolio standard). Not to minimize the accomplishment, but I should say that I hope so. If one of the most progressive and affluent communities in the U.S. can't get this done, I'm not sure who can. Their rate-payers are sufficiently enlightened (and wealthy) that they're willing and able to pay a slight premium for clean energy, and that's really all it takes, where the cost of solar and wind have come down to very near the point of grid-parity.

2) I'm glad to see that CSP (concentrated solar power) is still alive. I had a good talk with a VP at Abengoa, the Spain-headquartered solar thermal giant. Very smart guy, but not smart enough to know how to deal with a young

guy who joined our conversation and suggested that Abengoa adopt his concept of using supercritical CO_2 as a working fluid. If one of you chemistry/physics geniuses could comment on the pros and cons of this approach, I'd appreciate it.

3) Though there are a few technical people here, as the name of the show implies, most of them are bankers, lawyers, insurance folks, and other business people who, a few years ago, had never heard of renewable energy. As a result, the concept that we have a duty to move fast and hard in clean energy to combat climate change does not receive unanimous approval. I've run into several people who see renewables as a business with no more bearing on "doing the right thing" than selling hot dogs or greeting cards. I find that a bit strange. Yes, this is a business, but the world's a better place because it's here. I would think anyone involved would take a certain delight in that.

4) Apparently, there is some controversy about allowing the electric utilities to make money on their investments in energy efficiency. It certainly seems like the right thing to me. We need to create incentive for the utilities to sell less energy, not more.

5) I spoke with Dave Belote, a gentleman with an extremely powerful presence who had enjoyed a long and distinguished military career. He was the commander

of Nellis Air Force Base (sitting on 14,000 acres and employing over 12,000) before trading in his uniform for a suit, and working at senior levels in the Department of Defense. After his excellent talk on the U.S. military as a customer for renewable energy in which he referred to the fact that certain Congress-people are dead-set against this, I asked him what he thought laid behind this. He said, "There are few people in government who fairly and honestly look at the world in terms of what's best for the people. Almost all look through the lens of the interests of the people who got them elected." Needless to say, that's a tragedy.

6) The cost of an installed watt of PV is much higher in the U.S. than it is in Germany, as the Germans have made PV deals quite routine. In the U.S. huge legal fees are common, as each deal is minted uniquely. Apparently, American lawyers like things the way they are. One of the presenters went out on a limb with this statement, made, as it was, in a room filled with dozens of lawyers — at a conference sponsored by their firms, some of the largest on Earth:

The real problem with the solar market is that lawyers go out of their way to write documents that are deliberately ambiguous, just so they'll have the opportunity to argue about them and bill more hours. We don't have business negotiations any more; we bicker over legal language that shouldn't have existed in the first place.

Ouch. And here I thought *I* had a knack for insulting certain groups of people; this guy made me look like an amateur.

7) At Friday morning's "networking breakfast," we passed the microphone around each table, enabling each of the hundred-or-so participants to stand and talk for a minute or two about his/her background, purpose for attending, etc. Of course, I took the opportunity to promote 2GreenEnergy as a forum for identifying good ideas and taking them forward, uniting entrepreneurs with investors. I also told the audience what a huge supporter I am of ACORE's myth-busting website: EnergyFactCheck.Org.

8) I was reminded how common, nearly universal, stage fright is. There were no real meltdowns at the breakfast, but I noticed that the guy who spoke after me was trembling so violently that a moment after he had passed the microphone, he reached for a raspberry from his plate, tried to place it into his mouth, and nearly put it up his nose.

Again, good conference. I'm not a finance guy per se, but I always feel right at home at the show, and I certainly learn a great deal.

Film Project: Energy and Ethics

I'm in the process of making a five-minute video on "Ethics and Energy" to be entered in a contest later in the year. As we currently contemplate the project, the video will feature me giving a talk on the subject, supported by various "B-roll" clips and stills.

I rate myself "decent though far from excellent" at this type of public speaking. I'm certainly not good enough to deliver it extemporaneously without a bunch of flubs and omissions. So I just wrote a script for myself that I thought I'd share.

I need a more powerful opening – and closing. But here are the guts:

Human civilization is at a unique point in its history. Let me frame this with a few facts, at least according to what our scientists tell us:

Life on this planet is about 4 billion years old.

Homo sapiens, as a species, is somewhere between 100,000 and 200,000 years old.

What we call "civilization," i.e., people living together in permanent settings, began about 9,000 years ago.

In a period of 100 years beginning in 1950, the world's population will quintuple, from 2 billion to 10 billion, at the same time that the per capita energy consumption will go through the roof. Humankind will experience a huge increase in energy consumption in a minutely small period of time.

This energy gluttony is causing the depletion of natural resources – including some of the fossil fuels, like oil, that have been left to us from the distant past. Not only are we running out of some of those resources, but, in consuming them at this rate, we're rapidly ruining the only planet we have. We're polluting our oceans and our skies, we're poisoning our food with chemicals to keep up with demand, and our sea levels are rising due to climate change. We're converting the Amazon rain forest into farm land, largely for raising beef cattle, at the rate of 1.5 acres per second. In the time it will take you to watch this video, we will have lost an area the size of about three golf courses.

So, when I say that we're in a "unique position," what I really mean to say is that, for the first time in human history, we've begun to live in ways, largely in terms of energy consumption, that are literally unsustainable. What does this mean? "Sustainability" has all types of connotations: it's green, it's clean, it's eco-friendly. Maybe, I'd like to think, it's cool. But literally, something that is unsustainable is something that cannot be continued. And that's what we have here. Humankind will not be behaving this way in a few decades, even if it wants to, as it will encounter mass starvation, skyrocketing rates of disease, desertification of farm land, and tens of millions of "climate refugees," i.e., people fleeing from floods, and so forth.

So I'm suggesting that we have a huge ethical issue here, and that it has several different dimensions:

First, individuals need to take personal responsibility for the way they're behaving, in terms of the familiar phrase: reduce, re-use, and recycle. This has vast implications, of which here's a simple example: eating. If the whole world ate like the Chinese, where meat is used sparingly, almost like a decoration, the world could comfortably support about 15 billion people. But if the whole world ate like Americans, the food supply would only support about 2 billion.

Obviously, our eating habits are only one example. Think about the way we drive. In the United States, generally, everyone with a driver's license owns a 4000 pound car that has a single 150-pound occupant over 70% of the time it's on the road. This is grossly unaffordable, both financially and ecologically. And, as I say, this WILL change.

The other major ethical issue is that we're deliberately lied to about all this. Most of the people watching this video are either totally unaware of what I'm saying or think I'm some sort of communist hell-bent on destroying capitalism. And, by the way, no, I'm not. I ran a business with clients all over the world that at one point employed over 200 people.

The problem is that there are hugely powerful forces that are spending fortunes trying to convince you that there's nothing wrong with business as usual. Just keep on working, buying, consuming, and discarding whatever's left over. Drill, baby, drill. Your image of yourself DEMANDS that big car. Global warming is a hoax.

Speaking of human-caused global warming, a theory that's supported by over 98% of the climate scientists who study it, I urge the reader to look up an ad – a billboard – funded by the Heartland Institute, a conservative think-tank, whose members include most of the big players in energy, agribusiness, and the other mega-corporations that want to make sure that, as a consumer, you do not flinch. As you will see, the concept is to equate the mentality of a mass murderer with that of someone who's concerned about climate change.

Now personally, let's go back to ethics. What are we to make of multi-billionaires who want to make a few more billion dollars at the expense of the health and safety of the other seven billion people living on this planet – not to mention our grandchildren to come? What kind of human being runs ads to convince people to live in a way that is clearly destroying the planet? The word "evil" is not one that I throw around carelessly, but I honestly don't know what else to call this behavior.

So what to do? I guess the first thing is to come up to speed on the truth surrounding energy. Go to EnergyFactCheck.org, a website whose purpose is to separate the truth from the lies associated with renewable energy. Btw, these viewpoints that I've expressed here are my own, and they are not necessarily shared by the people who run that website; I just happen to think that it does a good job in debunking a great deal of the myths associated with the energy debate.

Also, if you're interested in a ton of free content on this subject, including a free newsletter, I would suggest you go to 2GreenEnergy.com.

We'll see how this goes.

Global Climate Change — Framing the Conversation

I had a chat with a friend the other day on the likely future for humankind vis-à-vis global climate change. He asked, "What do you think it will take, Craig, for us to muster the will to do something? If there were a fire burning in a building across the street, it would be apparent to everyone, and we'd do what we could to put it out. Here, we're talking about a set of inexactly known consequences that are unfolding over a period of decades. What will it take?"

Wow, that's a good question; I wish I had a good answer. All I can do is frame what I see as the big issues:

• Humankind applies a huge "discount rate" to situations like these; we place a much lower value on averting future pain than we do on enjoying current pleasure.

• It's not in mankind's DNA to be good at future planning, as my friend Tom Konrad points out. In the 100,000 years or so we have been here, if it worked last year, we do it again this year.

• Big Energy money is doing today what Big Tobacco did half a century ago, i.e., spending a fortune on misinforming voters and manipulating the political system in its favor. But eventually this will cease to work, as the lies become increasingly obvious.

• The analogy to tobacco is good, yet imperfect. Smoking, albeit an addiction, is something we can simply decide to live without. Our addiction to huge per capita uses of energy is not something we can break "cold turkey." Implementing energy efficiency, conservation, and renewables will require a level of commitment and willpower that doesn't seem to exist in any significant amount.

• Regarding meaningful intergovernmental agreements on GHG emissions, I think we're closer to international agreements eliminating all weapons, down to slingshots and pea-shooters. I.e., it won't happen. The COP meetings that focus on this appear to me a colossal waste of time, at least at this point in history.

• Might this change if there were a catastrophic event or concentrated set of events that science attributes to AGW (anthropogenic climate change)? It's possible, but, as I note often, we have astonishingly little regard for what scientists tell us. It would be interesting to see what happens if we lose the Greenland ice sheet and a few of the world's largest cities with it, due to the 23.6-foot increase in sea levels it would cause. Yet again, that's going to be a long and slow process, we seem to be pretty uninterested in things that aren't on fire right this moment.

I DO see a ton of great work being done at the level of individuals and groups, both large and small, in both the public and private sectors, in corporations as large as FedEx, and states the size of California. There are tons of incredible projects being embraced all over Europe and various parts of the rest of the world.

Is this the light at the end of the tunnel? I wish I knew.

Biomass Waste-To-Energy Solutions in France

Recently I was on a Skype call with a friend, a clean energy aficionado based in Central France. He went on at length about some of the projects he's pursuing, several of which are various types of pyrolysis plants that will invoke some extremely specific equipment that I never knew existed. I learned about machines that process corn and miscanthus, as well as:

• Plastic bottles, includes delabeling and unscrewing the caps (which have more calorific value per gram than the rest of the bottle). Who knew?

- Food waste, especially the byproducts from making chocolate that are particularly energy-rich. Again, news to me.

- Dying pines trees my friend believes were victims of climate change, many of which were planted by Napoleon's people who wanted to ensure France would always have enough wood to build war ships. Wow, a free lesson in European history!

We also talked about equipment to chill solar panels to create better efficiency and longer life. Can this be cost-effective? Apparently he thinks so.

As we were talking I had two main thoughts:

1) The Europeans sure are all over renewable energy solutions; this is a discussion you'd be unlikely to have in the U.S.

2) I don't see anything wrong with these concepts, and it's good that people are getting creative in their quest to reduce the consumption of fossil fuels — yet it's clear that none of these niche biomass ideas will make a strategic difference in the world energy picture. To me, the most exciting ideas are what I call "hitting the broad side of the energy barn," taking advantage of the fact that the Earth receives 6000 times more energy from the sun every day than all 7 billion of us use. We need a solution, or a set of solutions, that effectively harnesses 1/6000th of that energy, and we can all

go home and pretend we never heard of fossil fuels. Better yet, we can deploy efficiency and conservation technologies, making the fraction even tinier.

Having said that, I tip my beret to my friend, and to everyone doing anything they can to move in the direction of sustainability.

Heralding Corporate Sustainability Initiatives

I have recently been successful in lining up a few interns to help with the "corporate sustainability role models" blog that I've been chewing on for these last few months.

Here's the basic concept:

We want to herald the good things that are happening in the corporate world vis-à-vis sustainability – and there are a *ton* of them. Dr. Rajendra Pachauri (chairman of the IPCC) reminded me of this in our meeting last week (see interview earlier in this book). Where the 200+ sovereign countries of the world may be slow to come to agreements about climate change, many of the largest corporate entities are making fantastically large and completely sincere efforts to lower our ecologic impact on the planet.

Let's tell those stories. Let's interview spokespeople in as many of these companies as possible, and get a detailed explanation of the entity's past, present and future when it comes to corporate sustainability initiatives. This can take any of dozens of forms: products with better lifecycle analysis –

perhaps designed around biomimicry, business processes with lower carbon footprint, philanthropy in the "green" space – anything that contributes to better sustainability.

Of the many reasons I think it's important to tell these stories, perhaps the clearest is a basic sense of fairness. If you search 2GreenEnergy for the word "corporation," it's likely you'll either find a rant on an oil company's corrupt activities in Washington, or a reference to the scurrilous U.S. Supreme Court decision "Citizens United" in which corporations were given the right to spend as much money as they like in order to influence our elections. How about a bit of balance here? What about the many millions of corporate employees whose jobs are focused on changing the world for the better?

So far so good, but will people talk with us? Ha! I'm betting we bat close to 1,000. I can't imagine I'll have too much trouble getting a call returned whose purpose is to paint the guy and his company as super-heroes.

And here's a side benefit to our work here: A lot of these companies have active "venture" initiatives aimed at acquiring hot new technologies – some of which may be in cleantech. For example, I happen to know the president of "General Motors Ventures." Every once in a while, I'll come across an investment opportunity that I think would be a good fit, and I'll run it by him to see if his team may be interested. Now it's great that I happen to know one such guy, but I want to know *hundreds* of them – and these conversations will be an excellent way to get me there.

Is this a great idea or what?

Energy Policy – Hitting the Broad Side of the Barn

I sometimes wonder what I would do, given the power, to deal with climate change. I suppose the overarching principle I would use is prioritization. Why take on an issue that is contributing a microscopically small amount to climate change? Why not try to hit the broad side of the barn? I guess it will be cool when I can charge my cell phone using a solar photovoltaics (PV) fabric that's woven into my hat, but that's hardly going to change the world.

Here are four rough concepts that actually make a huge difference. Note that, in each case, the required technology already exists; there is no need to pull a rabbit out of a hat; in fact, my favored solutions are listed among the business opportunities on the 2GreenEnergy website.

1) Because people in developing nations have limited access to modern modes of generating energy, they tend to burn hydrocarbons, mostly wood and animal dung, for cooking and lighting; obviously, this contributes significantly to pollution in various forms. We need a micro-grid or off-grid solution like micro-wind, coupled with high-efficiency lighting, cooking and refrigeration. Fortunately, one already exists: WindStream. I'm trying to get my friends at the Eleos Foundation to invest in establishing a manufacturing facility in Kenya. This will provide numerous benefits all raveled into one: less poverty, better health and nutrition, and better education

(as people can read at night); note that educated people have fewer offspring. Everyone wins.

2) China is building a new coal-fired power plant at the rate of one per week, highlighting the world's need to reduce and ultimately eliminate this incredibly dirty form of energy. Yet, without a replacement, China (and most of the rest of the world) will simply continue to burn coal; it would be great if it were in short supply, but unfortunately, it's not. Solar PV and wind seem to be the two best candidates. PV is currently $0.56 per Watt – creating an attractive levelized cost of energy (LCOE) – especially if the issue of intermittency/storage can be addressed cost-effectively. This is a matter of R&D — and it's coming along very well; companies like Eos Energy Storage have announced breakthrough technology that have the potential to make this extremely affordable. Are there other things in our lives that store energy once the cost of batteries comes down? You bet: electric vehicles. Once this happens, it will mean the rapid phasing out of gasoline, diesel, and the internal combustion engine.

3) We continue to slash and burn the Amazon rainforests at the rate of 1.5 acres per second; in the time you'll spend reading this hort essay, we will have deforested an area approximately the size of a regulation 18-hole golf course. The driving force here is largely the need for additional pasture land for cows, enabling a growing population to have inexpensive hamburgers. Guess what? If I'm king

of the world, hamburgers will be getting a bit more expensive, because I'm putting an end to this deplorable behavior. Here, it is legislation, rather than technology, that is required.

4) Meat production aside, the energy footprint associated with the food we eat is outrageous. The average food item that Americans consume was trucked 1200 miles to reach our grocery stores. Moreover, there are "food deserts" all over the world, where the delivery of food is logistically so expensive and/or dangerous that it simply doesn't happen. Fortunately, we have an affordable solution to provide organic, locally grown food ready to go into place: Tower Harvest with its breakthrough in bioaeroponics. Again, I foresee a world in which factories spring up, employing thousands of people to build and sell these incredible, elegantly simple devices.

Clean Energy Business Plans for Your Consideration

The number of clean energy business plans I support reached 22 items at one point earlier in the year (though it's at 18 as of the date of publication of this book). Each of these, in my estimation, represents a real breakthrough that clearly has the potential to make a huge impact on the world energy scene. In particular, each carries with it a central question, the answer to which, I believe, is yes. For instance:

• Can microwind (anything under 5 kilowatts) prove cost-effective and reliable?

• Can a cost-related breakthrough in concentrated solar power turn the tide for this often-denigrated technology?

• Will a new approach in zinc-air batteries change the value proposition in electric transportation? Utility scale energy storage?

• Will ocean thermal prove to be an effective way to bring power to the one billion people living in the tropics?

• Can ocean current work effectively in places near the Gulf Stream, the Mozambique Current, etc?

• Will synthetic fuels prove to be workable in making use of off-peak energy?

• Has someone finally "cracked the code" with respect to waste-to-energy pyrolysis and gasification?

Again, the reason that these concepts are on the list is that I firmly believe in the efficacy of each. Now, is it possible that I'm wrong? Of course. Do investors need to conduct their own due diligence? Absolutely.

But I'll bet you know someone who wants a shot at changing the energy game, profiting greatly by helping the world migrate away from fossil fuels. Further, I'll bet that this "someone" will appreciate the work that I've done in reviewing well over 1200 concepts to arrive at these 18. If that's the case, I hope you'll let me know.

Sane Energy Policy Can Only Come From Honest Government

If we are to achieve a sustainable energy future, it will come as a result of our having found a way to create a viable working relationship between the public and private sectors. The private sector alone will not make any more than a token investment into an enterprise whose upside potential is purely long-term, as doing so will not achieve the demand for quarter-to-quarter earnings required to keep its management team employed. Moreover, the most powerful forces within the private sector stand to lose far more than they will win in the face of a migration from fossil fuels.

But how credible is it that government will participate in an honest and helpful manner? I hate to sound cynical, but I'm not a believer. Government has its own incentives and motivations that run counter to the public it ostensibly serves; its short election cycles mean political death for anyone brave, honest, or foolish enough to stand up for a long-term energy policy that will provide overall benefit to society.

It's also true that government in today's world operates unfettered by the checks and balances of old; one of the aspects of our current culture that frightens me most is how the media seems to go along with the most outrageous actions coming from the public sector. All around us, we see ridiculously shameful behavior that the media seems to either ignore or actively embrace. Are they asleep? Incompetent? Controlled? Some combination? I'm not sure.

Here are some top-of-mind items of the past couple of years that the media of 50 years ago would have ripped into like a hungry bear out of hibernation lights into a salmon:

• The executive branch's arrogation of powers clearly intended for legislative and judiciary.

• The U.S. Supreme Court's "Citizens United" decision, giving corporations the right to spend as much as they wish to influence our elections.

• Proposed legislation that would outlaw our military's ability to pursue its strategic interests in biofuels.

• Legislators who receive huge campaign contributions from the oil companies voting to continue federal subsidies to their paymasters.

- The 2012 National Defense Authorization Act which provides broad authority for the federal government to use the military in domestic operations in order to detain Americans indefinitely and without trial, nullifying the 4th Amendment to the U.S. Constitution, as well as the natural rights of Americans.

One's life goes by in the blink of an eye, doesn't it? It seems like only yesterday that we had sharp, brave reporters on the case, defending society against its most oppressive forces. The 20ᵗʰ Century was rife with great cases that will live in our memories, in which the media seemed to have saved us from runaway tyranny; in fact, I very firmly believed that they did in fact, do just that.

But let's take this discussion of "looking the other way" back a few thousand years. I'm reminded of a story I came across the other day about Tiberius, the emperor of Rome who reigned for 23 years after Augustus Caesar died in 14 CE. Apparently, the Roman senate thought a great deal of Tiberius, so much so that they sent him a message explaining that they would approve any measure he wanted to pass. But Tiberius, who may have been flattered by this, was quite troubled. He shot back a note to the effect that this was a very foolish concept. He asked, "What if I go crazy or for some other reason cease to make my decisions based on fairness and reason?"

But again, the senate re-issued its statement, encouraging Tiberius to offer up a wish list that they would immediately rubber-stamp. And again, Tiberius demurred.

After yet a third nearly identical round of this give and take, he sent a messenger onto the senate floor who, standing in the middle of the chamber, barked loudly, "Tiberius says: *How eager you are to be slaves!*" With that, the messenger immediately turned, and strode out.

Here in the 21st Century, we can pretend we don't notice the vast and growing corruption by which Big Money buys the political favors it needs to maintain its entrenched positions. We can ignore the cozy relationship between the government and the media that has fallen into its lap. In short, we can mind our own business (whatever that means). How eager we are to be slaves.

The Enemy of a Progressive Energy Strategy: Ignorance

I did an interview earlier this year for a consultant to the government of Israel who is amassing the viewpoints of folks in the clean energy industry. Apparently, Israel is on a mission to rid themselves of fossil fuels (especially oil, for obvious reasons) over a fairly short period of time, perhaps 10 years.

Bravo.

Of course, Israel's contribution to the world energy scene will be largely symbolic; it will scarcely move the needle on a planet with huge consumers like China and the U.S. Yet what a wonderful and important symbol it will be — a beacon of hope that humankind really *can* make a change in the way it generates and consumes energy.

In the interview, I mentioned that I find it very credible that Israel can make this happen; there are many factors that come together that make them one of the best candidates on Earth for such a rapid migration.

One such factor is its extremely well educated populace. Ignorance is the enemy of a progressive energy strategy. Here in the U.S., very few of us understand the implications of our coming into a room and flicking a switch, or walking into our garage and turning a key. For Americans, energy has always been there. And the few interruptions, e.g., the Arab oil embargo in 1973, have been sufficiently brief to convince almost all of us, young and old, that cheap, abundant energy is our right. We're soon to learn how incorrect that assumption is.

Climate Change and Energy Policy — Problems and Solutions

In connection with climate change and our energy policy, I recently heard someone say, "21st Century problems demand 21st Century solutions." I laughed derisively at the time, as it sounded like any one of hundreds of political slogans of yesteryear, like "Don't change horses in the middle of the stream," or perhaps a truism like "Plant your corn early." But I wonder if there isn't something more substantive here.

It's beginning to look like climate change may be the defining phenomenon of the 21st Century. By the year 2050, the effects of rising sea levels and extreme weather events,

Renewable Energy — Following the Money

coupled with a world population that will have exceeded 10 billion, could be causing hardship for nearly everyone on the planet in one way or another. Obviously the poor and middle class (to the degree to which that dwindling segment still exists) will suffer more than the rich, but it appears that no one will be having an easy time.

In terms of solutions, let's acknowledge that we are not powerless to do something about this. Having said that, it will take an unprecedented amount of effort and international cooperation. Countries around the world should be pouring enormous resources into clean energy R&D and energy efficiency. We all ought to be educating our people on the importance of energy conservation.

I suppose it's 100% correct to call this a 21st Century solution. It will be a kind of repurposing the political will of which we demonstrated ourselves capable in the 20th Century (remember what we did to win World War II? Put a man on the moon?), and combine that with technology that has advanced tremendously since that point.

Can we do it? Yes. *Will* we do it? I guess we'll see.

Drought, Extreme Weather, and Global Climate Change

Within a period of a few days earlier this year, I came across an article by frequent commenter Glenn Doty on the drought in the United States and physicist Dr. Richard Muller's op-ed in the New York Times, in which he summarizes the work performed over the past few years by the

Berkeley Earth Surface Temperature project. Until this point, Muller had been perhaps the most credible skeptic regarding studies connecting human activity to climate change; now, however, he concludes that "humans are almost entirely the cause" of global warming. "Call me a converted skeptic," he says.

Though the temperature of the Eastern Hemisphere (thank goodness) is not mirroring that of the U.S. this summer, most of us perceive the drought as an extreme weather event that is very likely the effect of climate change.

So what do we need? A worldwide energy policy, in which, among other things, we help China and the rest of the developing world to find sources of energy other than coal. It's true China put on a gigawatt of solar last year (in nameplate capacity, actual capacity about one-fourth of that). But they put on 40 gigawatts of coal, and will continue to add approximately one brand new gigawatt coal plant each week until someone can figure out another affordable path.

Not to sound pessimistic, but we're a heck of a long way from taking an active role here in the U.S., insofar as we face what I call the "ferocious five:"

1) A two-year election cycle that discourages leaders from committing to provide funding for R&D in this space.

2) A super-powerful lobby group controlled by the fossil fuel industry that has huge influence on our elections.

3) An electorate that is so concerned about its weekly paycheck that it pays little attention to national security, environmental damage, and other long-term, big-picture issues.

4) An electorate that is also receiving carefully structured propaganda from the energy industry to the effect that "global warming is the biggest hoax ever perpetrated on mankind."

5) A private sector that's still reeling from the financial collapse of 2008, managed by people compensated on quarterly profit, not the success of long-term investment.

I'm hoping we wake up, shake off our slumber, and *get to work*. I'm sitting right next to my office phone as I write this last paragraph. Maybe it will ring and president Obama will want me to come to Washington and serve as some sort of top advisor. On the other hand, maybe it won't.

Conclusion

Thanks for hanging in there with me through the peaks and valleys, the bumps and tight-twisting turns that made up this set of interviews and essays. Those who have been reading my books, reports, and blog posts over the past four years – or those who simply stumbled across "Following the Money" – can get a sense that this whole sustainability issue is far simpler than most people imagine. To sum it up with a few bullet points:

- What we've been told over the past 50 years – especially over the past decade – is true: humankind has come to a point at which "business as usual" will soon have terrible ecological consequences for everyone and everything living on Earth – now, and into the future. The world desperately needs an energy policy that brings together an aggressive combination of efficiency, conservation, and renewables.

- There is sufficient reason to believe that, if such a policy were it to be enacted even in part, it would bring with it a considerable rejuvenation in the sluggish world economy.

- As a species, we're supremely ill-equipped for this decision – or any others that affect our long-term survival. Whether, as some say, this is hard-wired into our DNA as a result of the evolutionary processes that got us here from our hunter-gatherer forefathers, or it's a part of the political evolution as groups and races formed and reformed over the past 30,000 years, I can't say. What I *can* say is that we're horrendously unprepared to come together, examine the growing body of scientific evidence, and make decisions that involve some mutual sacrifices based on the common good. The mere fact that I just typed out these words makes me something of a freak of nature.

- The inequality of the distribution of wealth and resources on this planet guarantees that all decisions of any real meaning will be made by a small handful of extremely powerful people. And keep in mind that these are people who, up until very recently at least, have seen nothing at all unsustainable about their business practices. What they do with the inescapable facts of global climate change, ocean acidification, loss of biodiversity, the ever-expanding damage to human health, etc., is largely up to them.

- Having said that, I have to think there is some import in what legendary environmentalist Paul Hawken says about the 200,000+ groups on the planet whose purpose is environmental and social justice. It's quite a notion to consider that 2GreenEnergy. com represents less than one-thousandth of one percent of the number of groups trying to affect change on this small but profoundly sick planet.

I suppose the real conclusion is that it's impossible to draw a conclusion at this point; it's up to each one of us to help forge a future that optimizes our chances to continue to thrive on our home planet.

As I'm fond of doing at the close of these projects, I'd like to offer my gratitude. I appreciate your reading my book. More importantly, all seven billion of us thank you for being among those trying to make a difference here while there's still time.

If you'd like to speak with me, please feel free to send me a note through 2GreenEnergy.com.

Craig Shields - 2013